Theology in the Age
of Scientific Reasoning

Cornell Studies in the Philosophy of Religion

EDITED BY WILLIAM P. ALSTON

Nancey Murphy

Theology in the Age of Scientific Reasoning

Cornell University Press, Ithaca and London

First published 1990 by Cornell University Press.
First printing, Cornell Paperbacks, 1993.
Second Printing 1996.

International Standard Book Number (cloth) 0-8014-2400-3
International Standard Book Number (paper) 0-8014-8114-7
Library of Congress Catalog Card Number 89-39375

Printed in the United States of America

*Librarians: Library of Congress cataloging information
appears on the last page of the book.*

♾ The paper in this book meets the minimum requirements of the
American National Standard for Information Sciences—Permanence of
Paper for Printed Library Materials, ANSI Z39.48-1984.

to
James William McClendon, Jr.

Contents

Preface

The best way to introduce this book may be to explain how I came to write it. The seed was planted during my years of graduate work at the University of California, Berkeley, where I had gone to study philosophy of science with Paul Feyerabend. Incidentally I learned there, for the first time, that Christian belief had fallen on hard times among the intelligentsia—a big surprise to a Nebraska ranch kid just emerged from the Catholic school system. It wasn't long before the question of the status of religious knowledge came to seem to me both more interesting and more pressing than that of scientific knowledge. The philosopher of science must answer the question "In what does the rationality of science consist?" Few besides my teacher Feyerabend would question whether science *is* rational. The philosopher of religion, on the other hand, must in these days provide an apologia for the very possibility of religious knowledge.

Philosophy of science has made great strides in this generation by careful study of science itself—as opposed to arm-chair investigation of the 'logic' of science. It seemed to me, therefore, that philosophers of religion needed an equally thorough knowledge of the cognitive aspects of religion. Thus upon completion of my work at Berkeley I immediately undertook a second doctorate, in theology, at the nearby Graduate Theological Union, where I concentrated on the history of theology and philosophy of religion.

During that time, the present volume began to take shape. I saw that Scripture, history, and the church's ongoing encounters with

God in community life and worship could be the data for a scientific theology. Theology itself (doctrine) could be accounted theories in a theological "research program." The analysis of the relations between data and theory, and criteria for acceptance of theories, would come from the philosophy of science of the late Imre Lakatos.

Lakatos put forward a revised version of a Kantian motto:

> Philosophy of science without history of science is empty; history of science without philosophy of science is blind.

I propose a motto that expresses a similar attitude regarding the relation between philosophy of religion and religion's content:

> Philosophy of religion without theology is empty; theology without philosophy of religion is blind.

Although this book is published in a philosophy of religion series, the reader will see that it is by no means empty of theology. I apologize to any theological readers who may find my recounting of the history tedious. Lakatos himself has been much criticized for failure to take history seriously enough.

The list of those to whom I am indebted has grown along with the book. First, I express my gratitude to Paul Feyerabend. I hope some day to emulate his thorough and creative scholarship.

I also thank the following friends and acquaintances who have read and commented on my manuscript or helped me with the project in other ways: William Alston, editor of this series, Ian Barbour, Donald Gelpi, Michael Goldberg, Philip Hefner, Gonzalo Munévar, Wolfhart Pannenberg, Arthur Peacocke, Ted Peters, Robert Russell, Kay Scheuer, senior manuscript editor at Cornell University Press, David Schultenover, Terrence Tilley, and Claude Welch.

My deepest gratitude goes to James William McClendon, Jr., who has advised and taught and encouraged me during my theological studies; who has read my manuscript several times, providing helpful comments and challenging criticisms; and above all, as my husband, has provided much needed support in many ways. It is to him that I dedicate this book.

NANCEY MURPHY

Kensington, California

Theology in the Age
of Scientific Reasoning

The Problem—Theological Method in the Age of Probable Reasoning

At the beginning of the nineteenth century the theological problem was, simply, "How is theology possible?" This was a question of both rationale and method, and included, at least implicitly, the question whether theology is possible at all. Of course, this had been a question in every age, but now it emerged with new strength and in a special configuration provided by the eighteenth century. The theologies of orthodoxy were still present, but fundamentally they were fighting rearguard actions as they retreated steadily before the forces of Enlightenment into the backwaters of intellectual and cultural isolation.

—CLAUDE WELCH, *Protestant Thought in the Nineteenth Century*, vol. 1

Friedrich Schleiermacher addressed his *Speeches*[1] to readers of Immanuel Kant. Today's cultured despisers of religion, however, are more likely to trace their lineage to David Hume—Kant having been 'refuted' by the discovery of non-Euclidean geometries and the replacement of Newtonian physics by relativity theory. Many theologians since Schleiermacher have been careful to take Kant into account, either by accepting his judgment regarding the limitations of theological knowledge (nineteenth-century liberals) or by turning his methods to their own purposes (early twentieth-century Lundensians, contemporary neo-Thomists). Yet, since Hume's sparring

1. *On Religion: Speeches to Its Cultured Despisers*, trans. John Oman from 3d ed., 1821 (New York: Harper and Row, 1958). First published in 1799.

partners died off, theologians have scarcely given him a thought, while the standard account among secular philosophers takes for granted his accomplishment—to have firmly barred the door to rational support of theism.[2]

Theologians' neglect of Hume may seem surprising, but perhaps it was just as well because, I shall argue, it is only with the epistemological resources of the past two decades that Hume's challenge can be met. This chapter will explore the demands Hume's arguments place on theology; the remainder of the book will show how these demands can be met with the resources of recent philosophy of science.

1. Theology and the Flight from Authority

There is wide consensus among philosophers today regarding the difference made to theological rationality by the Enlightenment, especially as incarnate in David Hume.[3] A survey of current philosophy of religion and its assessment of Hume's contribution would take another book. Fortunately, there appeared in 1981 a clear, historically cogent, philosophically and theologically sensitive work by Jeffrey Stout. Although some will disagree with Stout in detail, his *Flight from Authority* well represents prevalent views on Hume and

2. For example, I quote Wallace Matson's *New History of Philosophy*, vol. 2 (Orlando, Fla.: Harcourt Brace Jovanovich, 1987), p. 366: "Hume set out in his *Dialogues* not to demolish 'religion' altogether, but to accomplish a definite and limited task: to show that the inference from the alleged design in nature to an infinitely wise, powerful, and good Author of nature is invalid. He achieved his goal. It is true that William Paley (1743–1805), a generation after Hume, stated the design argument in its most popular form—the watch discovered on the desert island—and that in the years from 1833 to 1840 eight eminent British men of science published books under the auspices of the Royal Society expatiating on design in nature, for which service each scientist received a thousand pounds from the estate of the eighth Earl of Bridgewater. And versions of the argument are still found from time to time in *The Reader's Digest*. Nevertheless, few philosophers or even theologians in the past century have defended it. Hume was not the sole cause of the collapse; Darwin, in providing an alternative hypothesis to explain the curious adaptation of means to ends in nature, had enormous psychological effect. But for all that, Hume can be credited with what is rare in philosophy: a definitive refutation."

3. Antony Flew and Kai Nielsen represent mainline atheism in Britain and North America respectively. See, for example, Flew's *Presumption of Atheism* (New York: Harper and Row, 1976), p. 34; and Nielsen's *Philosophy and Atheism* (Buffalo, N.Y.: Prometheus Books, 1985), p. 18.

religion.[4] Therefore I propose to review this history as seen through Stout's eyes in the confidence that so doing will fairly represent theology's position today in rational discussion in our culture. Yet, if Stout's is the first word it will not be the last. Stout's updated Humeanism represents the challenge to be met in the remainder of this volume.

Stout first traces epistemological changes through the modern period (in philosophy, roughly, from 1650 to 1950), then shows their consequences for theology and ethics. In a preliminary way we may note that there have been two turning points in epistemology: The first was the rejection of the medieval concepts of knowledge based upon study of the authorities and deductive reasoning in favor of the modern period's foundationalism—that is, the concern with the reconstruction of knowledge on self-evident foundations (whether intuitionist or empirical). The second turn (still in progress) is the substitution of a holist approach for that very foundationalist doctrine—the holists having come to recognize that no sharp distinction can be drawn between one part of knowledge considered basic and some other part constructed upon it.

In effect, Stout argues that abandonment of the understanding of knowledge as legitimated by authority destroyed the rational credibility of theism. He sees the mainstream of theology since Hume as an assortment of unsuccessful strategies to protect theology against the consequences of this intellectual catastrophe. While criticizing theologians who argue that theology is, for one reason or another, immune from modern and contemporary rational standards, he is at the same time pessimistic about theism's prospects for support by these same standards of "probable reasoning."

1.1 Epistemology from Descartes to Quine

So two major epistemological changes separate us from Descartes: one was a change in available epistemic vocabulary that occurred shortly after Descartes's death; the other was the devastating criticism, 250 years later, of the foundationalist epistemological tradition that Descartes had initiated. There were two sets of epistemic vocabulary available to Descartes as a legacy from the Middle Ages and,

4. Notre Dame, Ind.: University of Notre Dame Press, 1981.

before that, from Aristotle. One set related to 'scientia' and had to do with demonstrative reasoning and certainty. The second set of terms was related to 'opinio,' which refered to all that falls short of demonstrative reasoning. Opinions must be judged more or less probable, and probability depends upon *approbation* of the proposition by the authorities, and in turn upon the *probity* of the authorities themselves.

Improbable though it may seem to us today, full-fledged notions of probable reasoning based on internal evidence came into European thought only in the second half of the seventeenth century—just after Descartes's death.[5] Lack of this modern notion of probability explains why Descartes took the option he did in responding to the skeptical crisis of his day. The crisis consisted in the simultaneous erosion of both of the epistemic categories available to him—both *scientia* and *opinio*. Trouble with *scientia* had begun in the late medieval period. The voluntarists' elevation of divine omnipotence and freedom threatened to narrow *scientia* to triviality: What could be deduced about the natural order if God could intervene to change it at any time? The only consequences one could rely on, it seemed, were those of the law of noncontradiction. Furthermore, the dissemination of ancient skeptical writings in the sixteenth and early seventeenth centuries gave rise to increasingly radical (modern) forms of skepticism.

Just as doubts about *scientia* placed a greater burden on *opinio*, the multiplication of authorities that occurred in conjunction with the Reformation made resort to authority an increasingly ineffective means for settling disputes. Descartes had three options. He could accept radical skepticism, or he could attempt to restore either *scientia* or *opinio*. He chose to attempt the restoration of *scientia*. His project, therefore, was to defend against skepticism by accepting in the beginning only those clear and distinct ideas that he could not doubt, and then to reconstruct all knowledge from them by means of deduction.

In years since, Descartes's intuitionism has been rejected—what is clear and distinct to one person is usually not clear and distinct to another in a different intellectual milieu. His reliance on demonstra-

5. Stout follows Ian Hacking's *Emergence of Probability* (Cambridge: Cambridge University Press, 1975). Another, perhaps complementary, account of the rise of empiricism involves the mathematization of science, with recognition only in our century that mathematical relations do not, strictly speaking, allow one to deduce conclusions about the natural world.

tion as the sole means of reasoning has also been abandoned as an impossible ideal, and later foundationalists have even been content with less-than-certain foundations. But what was not given up until the twentieth century was the foundationalist metaphor and the supposition that the only way to justify knowledge is to show that it is based on unproblematic foundations. More on foundationalism later.

Stout's explanation of the transition from medieval to modern epistemology is interesting enough to be worth summarizing here. Among the Jansenists at Port-Royal in the latter half of the seventeenth century there occurred a transformation in the sense of 'probable' (Latin: *probabilis*) from that of approbation by authority to that of the proportioning of one's belief to the weight of the internal evidence. The transformation happened more or less this way: If in the Middle Ages authority sanctions all beliefs except those that can be demonstrated from first principles, then surely the best authority one could have is God himself. God, according to Augustine and many who followed him, authored *two* books: the Bible and the Book of Nature. If nature is indeed a book and God its author, just as he is the author of the Bible, then events in nature, like linguistic expressions, are signs. To study them is to decipher God's meaning. Here natural observations do play a part in the determination of belief, but they do so *only because they are a kind of testimony* and therefore fit into the epistemic categories relating to authority. But soon, by means of the arguments of the Port-Royal logicians, the 'testimony' of nature came to have a new kind of status.

The Jansenist Blaise Pascal had already shown that the structure of reasoning about games of chance could be transferred to inference about other matters—recall "Pascal's wager." In Port-Royal's *Logic, or the Art of Thinking* (1662), the authors adapted Pascal's frequency calculations to create a "rule for the proper use of reason in determining when to accept human authority." This involved, among other things, judging the acceptability of an authority's pronouncement on a given matter on the basis of *frequency* of past reliability. Here we see one of our modern senses of 'probability' intertwined with the medieval sense.[6] Furthermore, if nature itself has testimony to give,

6. As will be apparent, my use of 'probable' in this book is intended to call to mind a longer tradition than might be recognized by current probability theorists. Inclusion (in Chapter 3) of Imre Lakatos, Thomas Kuhn, and others in the tradition of

then the testimony of a witness may be compared with the testimony nature has given in the past. Thus one may distinguish between internal and external facts pertaining to a witness's testimony to the occurrence of an event: external facts have to do with the witness's personal characteristics; internal facts have to do with the character of the event itself, that is, with the frequency of events of that sort. Given the "problem of many authorities" created by the events of the Reformation, the task increasingly became one of deciding which authorities could be believed, and the new sense of probability—of resorting to internal evidence—gradually came to predominate, making external evidence, the testimony of witnesses, count as evidence only at second remove. The transition from authority to internal evidence was complete.

So the followers of Descartes had at their disposal a new epistemic vocabulary—and with it new ways of answering the question of what it is rational to believe. The erosion of the ideal of *scientia* from that point on was no longer the catastrophe it had been for Descartes. In this sense, Hume stands on the near side of a great divide that separates us from Descartes—the invention of probable reasoning. But in another sense, Hume is more a student of Descartes than is often recognized when Descartes is considered the progenitor of modern rationalism and Hume of modern empiricism—for Hume continued Descartes's foundationalism, his only innovation being acceptance of a kind of foundational beliefs different from Descartes's clear and distinct intuitions.

The foundationalist argument may be summarized as follows: When we ask for justification of a belief, a chain of reasons that would constitute justification cannot be circular without begging the question and must not result in an infinite regress. Therefore the chain of reasons must at some point end in a 'foundation' needing no further justification. In Descartes's day deductive reasoning could be

------

probable reasoning reflects my judgment that these are the rightful successors of the neopositivists, whereas confirmation theory, with its reliance on probability calculus, is a dead end. I might instead have used the word 'empirical' but it has also been used in a variety of ways, most of which are too narrow for my purposes. I used the term 'scientific reasoning' in the title, but the change from an epistemology based on authority to one based on the proportioning of one's belief to the strength of the evidence was a much broader cultural change than that which occurred within science alone.

assumed to be the means of constructing the chain, once given the foundation. But when the foundations came to be thought of as sense experience, as already in Hume's day, the means of construction became problematic. In fact, one way of looking at Hume's famous problem of induction is to see it as simply drawing attention to the fact that because deduction only spells out consequences already implicit in the premises, there can be no deductive argument from a limited number of observations to a general conclusion. Yet nothing but deduction could provide the certainty that seemed a necessary condition for knowledge. Hume's solution was finally to accept the necessity of proportioning one's belief to the strength of the evidence.[7] Thus, in empiricist epistemology, both the nature of the foundations and the method of construction differ from the corresponding Cartesian elements, but the underlying assumption is the same—namely, that it is the business of philosophy to examine the justification of the putative knowledge claims of other disciplines by attempting to derive them from immediately given foundations.

Foundationalism has had its critics in the past (nineteenth-century Hegelians, for example), but since the middle of this century it has become a center of controversy. In 1951 W. V. O. Quine published "Two Dogmas of Empiricism," in which he argued (to the satisfaction of many) against the empiricists' doctrine of the analytic-synthetic distinction, showing that it could not be drawn precisely enough to make it useful, as well as against the dogma of reductionism—the belief that each meaningful statement is equivalent to some logical construct on terms that refer to immediate experience.[8] All at once the distinction long maintained between science (a purely empirical discipline founded firmly on sense experience) and philosophy (analysis of the meanings of concepts) was blurred.

Unwilling to use the 'modern' picture he had thus demolished, Quine suggested a model of knowledge different from that of traditional empiricism. The older view had seen knowledge as a layer cake. The bottom layer is sensory experience, which supports empirical generalizations, topped off by theories. (In Chapter 3 we shall see the steps by which this view of science was criticized and replaced

7. See *An Enquiry concerning Human Understanding*, ed. L. A. Selby-Bigge (Oxford: Clarendon Press, 1902), sec. vi "On Probability," and sec. x "Of Miracles," p. 110.

8. *Philosophical Review* 40 (1951), 20–43.

by philosophers of science as well.) Quine's new model pictures a belief system as a web or net. Beliefs that are most likely to be given up in the face of recalcitrant experience are located at the edges; beliefs less subject to revision are nearer the center. These latter beliefs are less subject to revision not only because they are further from experience, but also because they are interconnected with more elements in the rest of the system. When experience necessitates some change in the system, there are usually many ways to revise, including changing the meanings of some terms, revising theories, or even, Quine hazards, revising logic. The decision among these possibilities will in the end be pragmatic—how best to restore consistency with the least disturbance to the system as a whole.[9]

This Quinian model of knowledge, along with the epistemological and linguistic theses that go with it, is commonly referred to as holism. Holism rejects the distinction between analytic and synthetic truths because meanings can always be adjusted to fit beliefs to experience in preference to rejection of the beliefs. Holism denies that there is some class of basic beliefs that can provide an unquestionable court of appeal to settle disputes. Furthermore, it does away with the long-lived distinction between fact and value.[10]

Stout concludes in Part I of *The Flight from Authority* that once the holist picture of knowledge is accepted and the notion of meanings as fixed entities rejected, the mid-twentieth-century understanding of philosophy as conceptual analysis must change. Now the analysis of concepts can only be the description of how they are in fact used, and one will have to specify who uses them this way and when. Thus the natural successor to philosophy as analysis is a neo-Hegelian understanding of philosophy as history—"conceptual archaeology," in Stout's term. To find out what 'rational,' 'know,' 'justification' mean, one must see how the terms are used. In fact they are used

9. W. V. O. Quine and J. S. Ullian, *The Web of Belief* (New York: Random House, 1979).

10. Stout does not mean to claim, of course, that there are no more foundationalist epistemologists, only that with the qualifications now needed to evade criticism, there is no longer any point in holding a foundationalist view of knowledge. I agree with Stout that the new holist picture of knowledge offers greater promise. For more on holist epistemology and its relations to other disciplines, see Richard Rorty, *Philosophy and the Mirror of Nature* (Princeton: Princeton University Press, 1979); and Richard J. Bernstein, *Beyond Objectivism and Relativism: Science, Hermeneutics, and Praxis* (Philadelphia: University of Pennsylvania Press, 1983).

differently in different milieus—we have already seen that Descartes and his predecessors did not use '*scientia*' in the way we now use either 'science' or 'knowledge.' The historical events of the Reformation, the rise of science, Port-Royal, and other factors have all intervened to change the way these words function in relation to one another and to whatever else there is.

1.2 Theism after Authority

Stout's account of the fate of theism after the flight from authority is worth examining here, both for its analysis of Hume's pivotal role and because it offers theologians a view of their craft through the eyes of recent secular philosophers (at least of those who bother to think about it at all). Part II of *Flight* traces the changing fortunes of theism from Aquinas to Barth, showing the consequences for apologetics of the epistemological changes described above. Stout explains that although Aquinas and his contemporaries had good reason for accepting theism and, in fact, even for accepting "paradoxical" doctrines such as that of three persons in one God, these reasons, probably some time in the late seventeenth century, ceased to be good ones. Much of the history of theology and philosophy of religion since then can be read as attempts to come to terms with this fact. Regarding Aquinas, Stout says:

> I want now to reaffirm my claim that this reasoning justifies his acceptance of the "supernatural mysteries" of faith. The crucial point is that Aquinas accepted—and, unlike those who came several centuries later, had no compelling reason to abandon—epistemic principles that made the "supernatural mysteries" seem highly credible. The link between credibility and authority still firmly in place, Aquinas could accept on authority even the most "paradoxical" beliefs without violating any available norm of probable opinion. Authority makes opinions as probable as they could be. *Divine* authority confers objective certainty. Even a doctrine the very coherence of which must be taken on faith is perfectly acceptable by standards Aquinas had no compellingly good reason to revise or reject.[11]

Stout offers no opinion on the rationality of Descartes's theism, and such a judgment may not be possible if Descartes stands at the

11. *Flight*, p. 168.

dividing line between two different 'paradigms' of rationality. Contemporary philosophers are fond of giving Descartes's theistic arguments short shrift, but if we take seriously the historical nature of
rationality, such judgments may be anachronistic.

It is really the fate of theism after the Port-Royal *Logic*, and after
Descartes's death, that interests Stout. He describes deism as the
consequence of applying the new probable reasoning to religious
belief, distinguishing three stages of development seen most clearly
in Britain roughly from 1640 to Hume's death in 1776. In the first
stage, the argument from design was reformulated in such a way that
the order of the universe, its mechanical design, was taken as *evidence*
for a divine artisan. (Note the difference between this claim and that
of, say, Aquinas, for whom the purposiveness of agents in the world
is *proof* of a divine agent.) The concept of revelation was emphasized
in a new way, and miracles were taken as the means to relate revelation to probable reasoning. Stout takes his example from the work of
John Locke.

> For the most part, Locke works within the traditional distinction
> between demonstrative knowledge and the sphere of mere belief or
> opinion. The two grounds of assent short of the certainty of demon-
> strative knowledge are revelation and probability. Revelation takes
> precedence over probability in the sense that evident revelation re-
> quires assent even when the content of such revelation runs against its
> likeliness to be true. Reason is the guarantor of revelation in two
> senses. First of all, no true revelation could contradict the absolutely
> certain results of truly demonstrative knowledge. Second, reason
> must determine whether the putative revelation in question is indeed
> revelation, which must be done in accordance with probable reason-
> ing. If it is probable that the putative revelation in question is the work
> of God's hand, then the proposition commands our assent whether or
> not its *content* expresses a probable truth. One mark of divine author-
> ship, besides compatibility with what we already know and with what
> has already been certified as revelation, is the attendant presence of
> miraculous signs. Hence the importance of miracles for Locke.[12]

In a second stage of post–Port-Royal development, deists such as
John Toland and Matthew Tindal raised the question: if *all* evidence
is to be taken into account, as urged in Port-Royal's *Logic*, why

12. Ibid., pp. 115–16.

should not evidence for the *content* of putative revelation be taken into account as well? In fact, questions of the likelihood of the very idea of historical revelation were raised. Is it likely that a perfectly good God would have left the human race without decisive guidance for so long, only to grant the privilege finally to a small, isolated fraction of it? Is not the appeal to the "mystery" of revelation only an admission of the unintelligibility of the very idea? And even granting the inherent possibility of revelation, how likely is it that such a thing has actually taken place? Are there good grounds for believing in the actual occurrence of the miraculous events constituting the indispensable evidence for historical revelation?[13]

The important thing to notice, says Stout, is that the deistic response to the latitudinarian stress on revelation in effect unleashed internal (empirical) evidence from the carefully delimited place Locke had accorded it. Deism, in short, accepted only those tenets of traditional theology that could be established independently as probable hypotheses. In deism reason rendered revelation either improbable or redundant.[14]

Hume's critique of deism constituted a final stage of readjustment. First, he sharpened the deistic argument against using miracles to warrant belief in revelation. Miracles are by definition violations of the laws of nature. The laws of nature are established by uniform experience. If it is uniform experience that warrants belief in the probability of an event's occurrence, then belief in the violation of a law of nature (a miracle) is just what is least likely ever to be supported by good reasons. This argument was coupled with a low assessment of the credibility of the putative witnesses to miracles on the basis of the Port-Royal *Logic*'s criteria. Because the alleged witnesses are not known to be numerous, intelligent, well educated, evidently honest, in the position of having something to lose if the account proves false, and operating publicly in a well-known part of

13. Cf. Hans Frei, *The Eclipse of Biblical Narrative* (New Haven: Yale University Press, 1974), pp. 52–53.

14. Ronald Thiemann describes Locke's attempt to maintain both that the content of revelation is entirely in accord with reason and that revelation adds to the knowledge of God provided by reason as requiring an "impossibly fine balance." See *Revelation and Theology* (Notre Dame, Ind.: University of Notre Dame Press, 1985), pp. 17–24.

the world, we have especially good reason, Hume concluded, to disbelieve their accounts.[15]

Having thus intensified the deistic argument against revelation, Hume turned to the remaining core of deism itself—the argument from design. He pointed out that the hypothesis of an intelligent designer is only one possible explanation of the origin of the world and depends upon our first construing it as a machine or mechanism. If on the other hand we construe it as more analogous to an organism, then it could instead be produced by propagation. Because the hypothesis of an intelligent and morally good creator is not the only possibility, the final verdict must take into account all the relevant evidence, and the existence of evil in the world is prima facie strong evidence against this traditional conception of the creator.[16]

Thus Hume represents a great divide separating us from traditional theism, for in his work the consequences of the new probable reasoning were played out in theology. The burden of proof has shifted. Theology from Hume's day to the present seeks to defend itself not in the court of authority, but in the court of internal evidence.

1.3 Theology after Hume

The failure of theism to withstand the application of the new probable reasoning seems to give theologians a choice between two positions. Some accept Hume's critique as final and seek some other 'vindication' for religion and theology outside the cognitive domain. Others ignore the crisis created by the epistemological changes that ushered in modernity, and go on about their business as though Hume had never written.[17] The latter sort, however, pay the price of becoming intellectually isolated from and irrelevant to the host culture. Stout sees no third alternative. It is the thesis of the present

15. See "Of Miracles."

16. See *Dialogues concerning Natural Religion* in *David Hume on Religion*, ed. Richard Wollheim (Cleveland: World, 1963).

17. Many of those who go about their business as though Hume had never written do so not out of ignorance of Hume's works, but because they take Thomas Reid to have satisfactorily answered Hume. Reid's influence on American fundamentalist theologians was brought to bear by Princeton Theological Seminary professors John Witherspoon, Samuel Stanhope Smith, and Charles Hodge.

work that there is now a third alternative—a new possibility for
rational support of theism using the resources of postfoundationalist
(postmodern) epistemology—but before turning to it we should take
at least a brief look at the two unsatisfactory alternatives.[18]

According to Stout, the "dialectical progression" from Kant and
Schleiermacher to Hegel, Marx, Kierkegaard, Barth, and Tillich
takes Hume's accomplishment for granted. The attempt to refute
Hume on his own terms had retreated into obscure corners by the
end of the Victorian period.[19]

Kant and Schleiermacher set out to separate religious thought
from the realm of science, Kant removing it to the moral sphere and
the realm of "practical reason," Schleiermacher to the realm of feeling—an immediate awareness prior to the sort of discrimination
between subject and object that lies at the basis of scientific thinking.
Stout sees Hegel's accomplishment to be the reaffirmation of the
cognitive content of Christian doctrine. But Hegel's system would
fail on its own terms if the unities of thought in which he reconciled
the content of the Christian religion with the rest of culture failed to
represent real unities of life, and it did fail. Marx's criticism of the
Pietist spirit as withdrawal and Kierkegaard's diagnosis of the public
religion of the culture as essentially irreligious were conclusions to be
drawn from a careful reading of Hegel.

Stout endorses Barth's conclusion that there is no way around
Kant and Schleiermacher to the revival of natural theology and that
liberal theology is a dead end—a Schleiermacher will lead inevitably
to a Feuerbach. Barth is Stout's paradigm case of one of the two
variations of theism open to those who acknowledge Hume's accomplishment in that he abandoned apologetics altogether, claiming that
the revelation of God in Jesus Christ is the sole condition of whatever
'knowledge' we might have of him. On this view, theology can be
rational when it is returned to its proper basis in the authority of
God's word—theology is an attempt to seek partial understanding of

18. Because modern epistemology is so closely identified with foundationalism,
there is reason to suggest that the former ends with a thorough rejection of the latter.
Indeed I have argued that the modern period in philosophy came to an end at the
middle of this century with the rejection of foundationalism in epistemology, repre-
sentationalism in philosophy of language, and reductionism in metaphysics. See
below Chapter 6, sec. 3.

19. *Flight*, pp. 128–29.

what we believe because of grace—*fides quaerens intellectum*. Stout believes that Barth's insistence on the irreducibility of God's word carries Kant's isolation of theological discourse from the rest of culture to an extreme. In consequence, he undermines the preconditions for genuine debate with secular thought.

If Barth represents one variant of post–Humean theism, Tillich represents the other. Tillich keeps the terms of traditional theism but so alters their meaning that only the words remain. Stout follows Alasdair MacIntyre in designating this option as "giving the atheist less and less in which to disbelieve."[20]

Theology since Barth, Stout claims, is a sad story: "Those theologians who continue to seek a way between the horns, and thus to remain within the secular academy without abandoning the community of faith, have often been reduced to seemingly endless methodological foreplay."[21]

So, to sum up, the theologian's options, as Stout sees them, are: (1) ignore Hume, with the consequence that theology becomes irrelevant to the segment of the culture that has been affected by Enlightenment thought; or (2) take Hume's work for granted and either: (a) find some other vindication for theology (moral, aesthetic, existential), with the consequence that theology loses its cognitive content and becomes uninteresting; or (b) redefine terms so that theology has its own peculiar form of 'rationality,' with the consequence that theology becomes unintelligible to those who operate with the standard epistemology.

---

20. Ibid., p. 148. Cf. Nielsen's *Philosophy and Atheism*, pp. 20–21. Nielsen claims that atheism must be defined as the rejection of one or more of three sorts of theism: rejection of anthropomorphic concepts of God on the grounds that there (probably) is no such being; rejection of traditional concepts of God such as those of Luther, Calvin, Aquinas, Maimonides on the grounds that the concepts are incoherent; or rejection of some modern or contemporary theologians' concepts of God because they merely mask "an atheistic substance" such as love or moral ideals. Nielsen counts Tillich an atheist of the first type. See also Alvin Plantinga, "Reason and Belief in God," in *Faith and Rationality*, ed. Plantinga and Nicholas Wolterstorff (Notre Dame, Ind.: University of Notre Dame Press, 1983), pp. 16–93. On p. 16 Plantinga makes an obvious reference to Tillich, who he says has sought to replace belief in God with trust in the "Ground of Being." Among the "supersophisticates" who use the phrase 'belief in God' in such as way that to believe in God is not to hold any existential belief at all, Plantinga includes Rudolf Bultmann, R. B. Braithwaite, Gordon Kaufman, and John Hick.

21. *Flight*, p. 147.

I wish to endorse Stout's claim that Protestant theology has suffered a methodological crisis due to the collapse of the epistemology of authority. In fact in Chapter 4 we will see that this crisis struck Catholic theology as well, but nearly a century later. I also agree that *much* of modern theology employs in one way or another a 'defensive epistemology'—theologians seldom make straightforward claims to knowledge about God and God's relationship to the observable world. That the crisis is still with us is evidenced by the fact that many recent theological works focus on methodology or 'foundations,' and few theologians get past these prolegomena to the central (doctrinal and moral) business of theology. From an inside view, however, I believe the present crisis looks more like this: We are heirs to two major strands of theology—liberal and neo-orthodox. Insofar as no third option is discovered, theology remains in a state of crisis, for each strand leads to a dead end. For liberals there is the unanswered question of how *human* experience can tell us anything about *God*. For the neo-orthodox (and for fundamentalists as well) the unanswered question is how we can *know* with the required certainty that what we take to be revelation is indeed the word of God.[22]

I emphatically take issue, however, with Stout's pessimistic prognosis for theology, and I do so for two reasons. First, Stout has ignored at least one important theologian, Wolfhart Pannenberg, whose work represents the next step in the "dialectical progression" from Kant to Barth and Tillich. Pannenberg is a neo-Hegelian, reacting against both the liberal theology that preceded him and Barth's neo-orthodoxy. In fact, Pannenberg's own analysis of the present state of crisis in theology closely parallels Stout's. The next chapter examines Pannenberg's work and shows that despite his continental heritage, he does in fact meet Hume head on, defending the knowledge claims of theology in the court of probable reasoning. Thus Stout errs in concluding that vacuity or irrelevance have been the only options for theologians who take Hume seriously.

Unfortunately, however, I shall conclude that despite Pannenberg's impressive struggle against the Humean limitations to theological knowledge, he has not succeeded. Yet, by examining the point at which Pannenberg's system fails, we can discern the way

22. Cf. Thiemann, *Revelation and Theology*, p. 22. Thiemann traces this difficulty to Locke.

ahead more clearly. In brief, Pannenberg has employed an inadequate theory of scientific methodology, and this defect brings us to the main point of this book. Chapter 3 outlines what is arguably the best current account of scientific reasoning, that of Imre Lakatos. The remainder of the volume is dedicated to showing that Lakatos's methodology provides suitable guidance for the development of a theology that meets Hume's requirement—that beliefs about God stand up to the canons of probable reasoning.

Note that the important question is not whether theology can meet the requirements of the crude empiricism of Hume's day. Science itself could never have done so and cannot do so today. It is my assessment that theories of probable reasoning have become sophisticated enough only over the past two decades to allow for a meaningful test of scientific or theological knowledge claims. Thus I cannot be as scornful as Stout of the defenses developed by theologians during the years from Hume to Lakatos.

A second reason for disagreeing with Stout's gloomy prognosis is related to the first. Stout argues that the demise of foundationalism in epistemology makes no difference to the effectiveness of Hume's agnostic arguments. But close scrutiny of the new nonfoundational epistemology will reveal that it offers exciting possibilities for theological methodology. Again, this brings us to current philosophy of science, the source of the most detailed accounts of postmodern epistemology.

2. Meeting Hume's Challenge—A Strategy

The best demonstration that Hume has not foreclosed all possibilities for rational support of theism in this age of probable reasoning is to produce an instance of such support. Let us consider in the abstract what such a strategy might be. If we can place Hume in his own context so as to see the problems and arguments to which he responded not as *the* timeless questions of theism, but as particular, historically conditioned apologetic moves, then we may also see that philosophy of religion can escape the shadow of his negative conclusions. Also, we must take into account the difference between Hume's understanding and ours of such matters as what counts as adequate evidence. My claim is that when we consider (1) the fact that Hume's defeating of two apologetic strategies was not equivalent to the defeat of theism itself, and (2) the vast changes since Hume's day in our understanding of how evidence is brought to bear to support theories, we will see that Hume's agnostic arguments, although devastating in their day, have lost their teeth with the passage of time.

Hume's strategy in his attack on apologetics was to divide and conquer. This was a particularly apt strategy because theists themselves were divided into two camps, each with its own apologetics. In fact, the deists with their argument from design were aligned with Hume in attacking the more orthodox theists' attempts to validate revelation on the basis of miracles. Hume's strategy is often described as dialectic: in *The Natural History of Religion* (1757) the claims of orthodoxy were made to rest upon rationality alone, and then this rationality was subjected to devastating analysis in the *Dialogues concerning Natural Religion* (1779).[23] Richard Wollheim more accurately, I believe, describes Hume's attack as threefold: In the *Dialogues* he showed the ineffectiveness of rational arguments for the existence of a deity, either a priori or a posteriori. In *The Natural History* he showed the ineffectiveness of any appeal to historical tradition to support religious truth. Finally in "Of Miracles" he attacked the ultimate refuge of religion—revelation.[24]

Two possibilities for circumventing Hume's arguments are immediately apparent. One is to find a different means of validating revelation that does not rely on belief in miracles. This was indeed the ordinary apologetic in Hume's day and before, but it is obviously not the only possiblity. A look at theology today shows that numerous alternatives have been developed.

Second, it bears repeating that the weakness of the theistic arguments Hume considered and so easily demolished in the *Dialogues* resulted from the separation of revealed knowledge of God's nature from an abstract concept of God that was then to be supported by evidence from experience. For the deists of Hume's day, this made *order* the only relevant evidence for God's existence and nature. As

23. See, for example, Antony Flew, "The Impossibility of the Miraculous," in *Hume's Philosophy of Religion* (Winston-Salem, N.C.: Wake Forest University Press, 1986), p. 10.

24. See Wollheim's Introduction to *David Hume on Religion*.

Hume showed so skillfully, such an argument standing alone is too weak and one-dimensional to be either convincing of God's existence or informative about his nature.

Thus it appears that the crucial step for an effective apologetic strategy must be an approach to theology that does not distinguish between natural theology and theology of revelation, and one that draws upon religious tradition to provide a nuanced concept of God. This approach will be expected, in turn, to make relevant a greater variety of evidential relations, and from this point on it becomes necessary to have a more complex *theory* of evidential relations—it calls for the best of the developments in philosophy of science since Hume's day. In the end we may see that (contra Stout) Hume's arguments *do* fail, partly because of the limitations inherent in his empiricist foundationalism. Thus we shall have to take issue with Stout's (and others') claims that epistemological changes since Hume's day have not undermined the latter's position.[25] The emergence of holism and its consequences in philosophy of science have drastically changed what it means to supply evidence for a hypothesis. What could not be done with Hume's and others' simplistic notions of evidence can be done today with a more adequate view. So holism per se is not the issue, but holism must be viewed as of a piece with other changes in both theology and philosophy of science.

Let us now look at Pannenberg's proposal for weaving methodology, tradition, doctrine of revelation, and concept of God into a "scientific" theology.

25. For example, see Nielsen, *Philosophy and Atheism*, n. 6 to text on p. 18.

The Quest for Theological Method—Pannenberg versus Hume

Theologians—despite their attempts to reformulate Christian theism in terms acceptable to a secular academic audience—continue to skirt the issues implicit in Hume's dictum that a wise man proportions his belief to the evidence.

—Jeffrey Stout, *The Flight from Authority*

If the majority of recent theologians have merited Jeffrey Stout's harsh criticism, Wolfhart Pannenberg is one notable exception. Pannenberg's may be the boldest attempt in this generation to establish the credibility of Christian belief vis-à-vis the canons of probable reasoning. Furthermore, Pannenberg's system provides an almost point-by-point counter to the positions argued by David Hume, despite the appearance of vast differences in philosophical presuppositions—British empiricist versus continental Hegelian.

The strategy for circumventing Hume's negative conclusions sketched at the end of the preceding chapter included a doctrine of revelation validated on some basis other than miracles, a concept of God drawn from religious tradition rather than from philosophical speculation, and a new (more complex) account of relations between theory and evidence. Pannenberg's system incorporates all these moves. To see how it does so, we must consider four interrelated elements of his theological program: (1) his doctrine of revelation, (2) his concept of God, (3) his methodology and its relation to scientific method, and (4) his proposed order of argument for the existence of God—or put differently, his plan for a scientific theology.

Even so, it will in the end appear that Pannenberg, too, is defeated by the arguments of the Scottish philosopher of two centuries past. Yet, because Pannenberg's system fails for want of an adequate theory of scientific method, it can be repaired if we can reconstruct it in light of a more adequate philosophy of science. This chapter will be dedicated to explication and criticism of Pannenberg's system; in Chapter 6 I shall indicate how his system can be recast as a scientific research program and thereby evade Humean objections.

1. Pannenberg's Theory of Revelation

Pannenberg objects to those twentieth-century theories of revelation that in one way or another make Christian belief immune from criticism. He accuses Karl Barth of "scriptural positivism," in that no answer can be given, on Barth's view, to the question how we can know that *this* is God's word—revelation is supposed to be self-authenticating.[1] While Pannenberg agrees with the dialectical theologians in emphasizing that the content of revelation is God himself, he maintains that we do not encounter God directly.[2] We come to know God indirectly by means of his acts in history. Although the Christian theologian's interest will focus on certain events—the history of Israel, Jesus, and the church—Pannenberg believes that it is a mistake to select a special salvation history apart from universal history, immune from historical critical methods and interpretable only from the point of view of faith.[3] Following the German idealist tradition, he proposes a theory of revelation as history: God is revealed in the whole of the historical process, a process open to inspection by believer and unbeliever alike.

1. See, for example, Barth's *Church Dogmatics*, I/1 (Edinburgh: T. and T. Clark, 1936); and his *Epistle to the Romans* (London: Oxford University Press, 1968).

2. Pannenberg's theory of revelation is set out in *Revelation as History*, coauthored with Rolf Rendtorff, Ulrich Wilckens, and Trutz Rendtorff (New York: Macmillan, 1968). Trans. by David Granskou of *Offenbarung als Geschichte* (Göttingen: Vandenhoeck und Ruprecht, 1963).

3. See, for example, Oscar Cullmann, *Salvation as History* (London: SCM Press, 1967); or H. Richard Niebuhr, *The Meaning of Revelation* (New York: Macmillan, 1941).

But the idealists' theory of revelation-as-history has had its own besetting problem: to explain how a single event in history (the "Christ event") can have absolute meaning, because it would seem that no single act of God could reveal him completely—only the totality of his acts could do so. Yet the totality of history is not available for inspection. Pannenberg claims that this problem can be solved by attending to the *eschatological* nature of Jesus' resurrection—it is a foretaste of the end of human history. In a limited sense we know ahead of time how the entire course of history will come out: the end will be on a cosmic scale what has already happened in Jesus. The God of Israel, of whom we claim to have knowledge based upon his actions in the past, will be confirmed as the God of all people on the last day. Thus, while it *is* only the whole of history that demonstrates the deity of the one God, there is still one particular event that has absolute meaning as the revelation of God insofar as it anticipates the end of history.

Pannenberg's understanding of the nature of revelation leads to a corresponding account of the role of Scripture in Christian thought. Scripture witnesses to God's revelation insofar as it foretells or reports God's historical acts. He associates his understanding of Scripture with Gerhard von Rad's view of the texts as products of the transmission of traditions. The traditions are constantly reinterpreted in light of later events until we come to the New Testament where Jesus' resurrection is (in the sense just explained) the *last* event, and all of history to that point (as recorded in the Old Testament) is given its final reinterpretation from that perspective. This is the sense in which the New Testament has final authority. No 'later' events can now occasion further reinterpretation.

In contrast to many theologians who pay scant attention to apocalyptic literature, Pannenberg sees it as the final step by which the Old Testament writers appropriated the whole of history for JHWH's dominion and as a necessary presupposition for understanding Jesus' resurrection as God's definitive act of self-revelation. Pannenberg and his coauthor Ulrich Wilckens argue that Jesus' own overarching presuppositions were apocalyptic. Jesus' understanding of himself, of God, and of history—indeed his entire message in word and deed—is revelatory because the whole of his life was vindicated by God's raising him from death.

2. Pannenberg's Theological Method

My exposition of Pannenberg's theory of revelation-as-history leads directly to the question of the nature of theological method. If theology takes revelation as its norm, and if revelation occurs in and through historical events as recorded in Scripture, then theological method must be historical method. Yet this cannot be the whole of it. Pannenberg emphasizes that the very meaning of the word 'God' requires that theology be more than interpretation of the Scriptures. He adopts Rudolf Bultmann's definition, intended to be more or less tradition-neutral, that God is the all-determining reality—the power that determines everything that exists. But this philosophical conception of God must be corrected and transformed in Christian theology by its application to the God of Israel.

As a result of Pannenberg's dual concept of God, Christian theology must take place "within the tension between two tendencies": one is the concern about the faithfulness of theology to its origin in the revelation of God in Jesus Christ as witnessed in the Scriptures. On the other hand, theology must be related to all truth whatever because all being must be understood in relation to God, the "all-determining reality." Ultimately, the theologian must attempt to relate all other knowledge to the God of the Bible and to attain a new understanding of everything by viewing it in this light.[4]

Here Pannenberg deliberately rejects the tendency of modern theology, epitomized by Barth and criticized by Stout, to isolate itself from the rest of knowledge. When theology limits itself to the task of interpreting Scripture, it gains the advantage of peaceful coexistence with the other faculties of the university. Yet the price of conceiving of theology as the science of revelation is estrangement from and irrelevance to the secular sciences.

Pannenberg's definition of God and doctrine of revelation make it necessary to discuss his theological method under two headings: (1) the use of Scripture in theology—historical and hermeneutical methods, and (2) confirmation of the Christian tradition vis-à-vis all known reality.

4. Wolfhart Pannenberg, *Basic Questions in Theology*, vol. 1 (London: SCM Press, 1970), p. 1. Trans. by George H. Kehm of *Grundfragen systematischer Theologie, Gesammelte Aufsätze* (Göttingen: Vandenhoeck und Ruprecht, 1967), p. 7. Hereafter *BQT* and *GsT* respectively.

2.1 The Role of Scripture in Theology

In Luther's day it was possible to hold the doctrine of the clarity of Scripture—that its essential content arises clearly and univocally from its words. Hence Luther could claim that his doctrinal statements were simply restatements of the contents of the New Testament in a different form. Post-Enlightenment theology recognizes two problems of distance that were not problems for Luther: (1) the difference between the literal sense of the texts and their historical content—that is, the distance between the texts and the events they record; and (2) the historical distance between the theologian today and the texts of the primitive Christian period.

It is common to take the distance between text and the contemporary interpreter as the central problem for theological methodology. The theory of hermeneutics deals with the problem of the "repetition" of the same content in a completely changed situation. H. G. Gadamer describes the task of spanning the distance between text and interpreter as that of the "fusion of horizons." But *how* this can be done, Pannenberg notes, is an open question in theology today. Pannenberg answers that we are able to understand Scriptural texts because we already share a common horizon with the primitive Christian church, simply by virtue of the fact that we belong to a common historical process. *Recognition* of this common horizon requires two things: first, that it be shown to the secular present that its own hope for the future can be realized in the foundational event of primitive Christianity; and second, that we have the idea of universal history—that is, that all of history is moving toward a common goal. Such a notion is our legacy from Israel via the Christian church.[5] Put more briefly, what is required in order to make the New Testament texts speak meaningfully today is a recognition that the resurrection of Jesus answers for all of us, despite our far remove from first-century Palestine, the questions of where history is going and of our own personal fate.

Thus the question of what the Scriptural texts *mean* today is integrally related to the question of what really happened. If in fact the revelation of God takes place in historical events and is available to anyone who witnesses those events then, given our shared horizon of

5. *BQT*, pp. 9–12; *GsT*, pp. 17–20.

universal history and hope for the future, what is required to make the texts speak meaningfully today is to uncover and clarify the events themselves. This is the task of the historian; theology is hermeneutical, but its hermeneutic *method* is historical investigation. The task, however, is not merely to compile a catalog of neutral events (*Historie*), but rather to assign events their place in a meaningful whole.

Pannenberg notes that ultimately the historian must assume some ground for the unity of the totality of history, with regard to which all events take on their ultimate significance, and only the concept of God can provide such a ground. It is the concept of the *one, faithful* God who is both creator and eschatological judge that renders all events a single unified whole. Notice, though, that even if the historian is with varying degrees of consciousness assuming a transcendent ground of unity, the theologian's emphasis or direction of inquiry differs from that of the historian. Whereas the historian *assumes* a "pre-projected unity of history," the theologian must *inquire* specifically about this universal context of meaning. Of course the theologian's projection of the unity of history is, as is the historian's, in need of confirmation.[6] Among all possible confirmatory events, Christians' theological interest favors those associated with Jesus of Nazareth.

So we return, after distinguishing the nature of historical and theological investigations, to the more concrete question about the means for investigating the events surrounding the life and death of Jesus. The controversy regarding what can and cannot be known about Jesus from historical investigation has been fierce and long-lived. Pannenberg enters the discussion not at the level of particular findings, but by challenging methodological decisions made by historians, beginning with Giambattista Vico (1668–1744), which systematically rule out the possibility of certain findings regarding New Testament events. It is specifically the anthropocentrism of modern historical-critical method that Pannenberg calls into question, because it has two unacceptable consequences. First, by pinning the unity of history to the human story (rather than to God's) it fragments history and results in relativism. Second, it is apt to exclude all transcendent reality as a matter of course. A principle of analogy—that putative past events are to be ascribed probability on the basis of

6. BQT, p. 199; GsT, pp. 171–72.

resemblance to current events—is part and parcel of this anthropo-centric methodology. Pannenberg claims, however, that it is not knowledge of regularity (and thus of abstractions) that constitutes historical knowledge, but rather that history is knowledge of (concrete) particulars. The recognition of analogies can be used in a limited way to allow for discovery of the *limitations* of what is held in common between ourselves and the past—to see the differences along with the similarities. Given this restricted role of analogy, we cannot deny knowledge of that which has no exact analogy. In the case of a genuinely unique event, however, the historian's knowledge is limited. While it can be known *that* such an event has taken place, the event still remains "opaque"; it is not *understood* in the way a familiar event is understood.[7]

There is a second role for analogy in Pannenberg's conception of historical method. In the study of historical documents, including the Scriptures, exact analogy to a form of tradition that has no referent (myths, legends, and the like) is good reason for concluding that the reported 'event' is not historical.

In sum, though not denying the importance of analogy with the present for our understanding of the past, Pannenberg does deny that knowledge of the past is strictly limited to events of the same type as those in current experience, for if historical-critical method were to take analogy to common current events as a necessary condition for establishing the occurrence of a past event, then it would systematically exclude the possibility of any knowledge of God's *unique* actions in history.

Because analogy to current events is not the powerful criterion for Pannenberg that it is for other historians, he must supply additional means for deciding on the historicity of events. We will see in the next section that his method is much like that for testing any scientific hypothesis.

2.2 Confirming the Christian Tradition vis-à-vis All Known Reality

Pannenberg claims that if theology is only a "science of revelation," based on the Scriptures alone and not confirmed by present experience, then it is no science at all. Yet, as I emphasized above,

7. *BQT*, pp. 35–48; *GsT*, pp. 43–54.

confirmation of the existence of God must make use of religious tradition.

Pannenberg agrees with Stout regarding the source of the question of the scientific status of theology. Since the Middle Ages theology has *claimed* to be a science; can it now, after the age of authority, defend its knowledge claims using generally accepted epistemic standards?

> The . . . concern, to defend the truth of Christianity by generally accepted criteria, has been present since the thirteenth century in the argument about the scientific status of theology and its right to be included among the sciences taught in a university. If theology were now forced to disappear from the universities on the ground, maintained by many people, that it is essentially tied to authority and therefore unscientific, this would be a severe setback for the Christian understanding of truth. . . .
>
> The questioning of the scientific character of theology within theology itself is paralleled in recent discussion in philosophy of science by influential tendencies which seek to deny Christian theology any claim to scientific validity. As a reaction to the efforts of many theologians to provide theology with general immunity from rational criticism, these tendencies are largely understandable. They have, however, the additional effect of helping tie theology to a justification of its thematic which is irrationalist and based on an appeal to authority, even though this in turn constitutes another ground for criticism of theology.[8]

Pannenberg recognizes that the paradigm of science today is natural science, which no longer needs epistemological justification but instead, because of its success, prescribes where and in what sense we may talk of knowledge. Epistemology has been replaced by philosophy of science.[9] Consequently he begins his case for the scientific status of theology with a history of philosophy of science from logical positivism through the critical rationalism of Karl Popper.[10] While logical positivism held sway, there were apparently two

8. Wolfhart Pannenberg, *Theology and the Philosophy of Science* (Philadelphia: Westminster Press, 1976), pp. 13, 20. Trans. by Francis McDonagh of *Wissenschaftstheorie und Theologie* (Frankfurt a.M.: Suhrkamp, 1973), pp. 17, 24. Hereafter *TPS*; *WT* for the German edition.

9. *TPS*, pp. 26–27; *WT*, p. 28.

10. See Popper's *Logik der Forschung* (Vienna, 1935), trans. Popper et al. as *The Logic of Scientific Discovery* (New York: Harper, 1965).

routes for theology to take: to deny the cognitive meaning of theological statements or to criticize the empiricist criterion of meaning.
Pannenberg gives Popper credit for recognizing the difficulties of
logical positivism and thus, indirectly, freeing theology from its
restrictions. But there remains the question whether Popper's methodology of falsificationism is the proper criterion by which to judge
the scientific status of theology. Pannenberg concludes for several
reasons that it is not: First he notes the holist criticisms of Popper's
methodology, raised especially by Thomas Kuhn, and then argues
that neither history nor the historical elements in the natural sciences
can be treated methodologically as Popper's views would require.[11]

Despite his rejection of Popper's criterion for scientific reasoning,
Pannenberg endorses his position regarding the *anticipatory* character
of all knowledge. In Popper's view, the basic statements upon which
science relies are not incorrigible—they are accepted on convention
by the scientific community but can always be called into question by
later discoveries and subjected to testing and perhaps revision. Thus
scientific knowledge is based on the anticipation that further testing
will continue to support one's conclusions. But even low-level empirical statements are anticipatory—dependent upon predictions that
future sensory experience will not call them into question.

Having rejected Popper's methodology of falsificationism, Pannenberg proposes that theories are to be criticized on the basis of how
well (coherently, parsimoniously, and accurately) they account for
all the available data. This view takes account of (some of) the
criticism brought to bear against Popper's methodology and has the
further advantage of being general enough to apply to history and
even to philosophy.[12]

Pannenberg next addresses the relation between the natural sciences and the human sciences. He follows Wilhelm Dilthey (1833–
1911) in maintaining that the distinctive task of the human sciences is
the understanding of meaning, which is brought about by placing
human phenomena as parts within the larger 'whole' to which they
belong. This hermeneutic process, according to Pannenberg, always
involves at least implicitly a conception of the whole historical pro-

---

11. See Thomas Kuhn, *The Structure of Scientific Revolutions*, 2d ed. (Chicago:
University of Chicago Press, 1970).
12. *TPS*, chap. 1.

cess, because the meanings of events change as history progresses.
Understanding the meaning of human action is always dependent on
"semantic constructions which in turn are related to the totality of
meaning which constitutes the horizon of any given human group-
ing's experience."[13] This being the general method of the human
sciences, Pannenberg argues, there is no real difference between the
methods of the human and natural sciences—explanation in both can
be understood in exactly the same terms.

> The explanation puts forward a new frame of reference within which
> the previously unintelligible event now becomes intelligible. A frame
> of reference of this sort can be set up by a hypothetical law, but also by
> other things. Stephen Toulmin has described the main effect of expla-
> nation as 'making sense' of previously unintelligible observations. In
> Toulmin's view this takes place not only through the construction of
> hypotheses but also quite generally in the construction of 'ideals of the
> natural order' which are used by human beings as 'explanatory para-
> digms . . . to make nature intelligible to them.' This remark implies,
> even though Toulmin does not stress it explicitly, a starting-point in
> systems-theory rather than in a merely nomological notion of expla-
> nation.[14]

Thus Toulmin's understanding can be applied to all types of explanation: they all function by placing the fact to be explained in a
context in which it can be understood as meaningful. In history the
context is the series to which the event belongs. In natural science a
single event is explained by being shown to be an instance of a natural
law, and the law itself is explained by being situated within the
context of a theory, and the theory within the context of an ideal of
natural order. In hermeneutic method the text is understood by being
related to the semantic whole to which it belongs.

With these conclusions regarding the nature of scientific method,
Pannenberg turns to the question of the scientific status of theology.
Two recent views on the matter are those of theology as the scientific
study of religion and as the science of God and of revelation. Pannenberg attributes the latter view to Barth, whom he commends for
returning theology's focus to its proper object, namely God. On the
other hand, he criticizes Barth for basing theology on faith, which is

---

13. *TPS*, pp. 102–3; *WT*, p. 104.
14. *TPS*, p. 139; *WT*, pp. 140–41. References to Toulmin's work are from
Foresight and Understanding (New York: Harper, 1961).

not available to critical reason, and thus making dialogue with those outside faith impossible. The problem for a scientific theology, as Pannenberg sees it, is to show that God's existence and dominion over history can, through reflection on the Christian tradition, be substantiated in a way that does not depend on an "arbitrary venture of faith on the part of the theologian."[15]

Theology is a true science of God (and also a true science of religion) when it investigates religious traditions to see to what extent their conceptions of the whole of reality are able to take account of all currently accessible aspects of reality. That is, traditional claims of religions are to be regarded by the theologian as hypotheses to be judged on the basis of their ability to integrate the complexity of current experience into their religious understanding. The criteria (from Pannenberg's earlier discussion) are parsimony, coherence, and accuracy.

To sum up, I began with Pannenberg's view that Christian theology takes place "between two tendencies." We can now see precisely what this means. On the one hand the Christian theologian is responsible for studying the Christian tradition in order faithfully to understand what it has to say about God. This is done by means of the hermeneutical-historical method, which itself raises the question of the totality of the historical process and the source of its unity. On the other hand the theologian is responsible for testing the Christian tradition's concept of God and of the 'whole' against everything that is known today about the course of history, and for judging whether the tradition's understanding is the best available account of the whole of experience.

The theologian's task becomes scientific because of this second aspect. As we saw above, Pannenberg conceives the task of the scientist to be the proposal and testing of general theories on the basis of their capacity to make intelligible a wide variety of phenomena. Theology's method is exactly the same as that of the scientist (whether natural or human science); its only difference is the level of generality. Theology deals with the broadest possible context—with theories about the source of unity of the whole of (historical) reality. These theories arise from religious experience but are subject to testing against all other experience.

Pannenberg's discussion of theological method is complex and at

15. *TPS*, p. 277; *WT*, p. 279.

times unclear. Let us therefore return to the question of Jesus' resurrection to see how he actually supports his position. This examination will also give us a starting point for an overview of Pannenberg's program for scientific support for the truth of Christian theology.

3. Pannenberg's Scientific Theology

Jesus' resurrection plays a crucial role in Pannenberg's theology, because it solves the problem of how we can know ahead of time about the totality of history. This problem arose from Pannenberg's Hegelian doctrine of revelation, and we can now see that it arises also from his general theory of method: all understanding implicitly raises the question of the totality of history, and theology in particular seeks to evaluate theories about this totality. Pannenberg claims that in Jesus' resurrection we see in microcosm the transformation that awaits the entire cosmos, and in this the vindication of the God of Jesus as the all-determining reality. With so much hanging on this peculiar event, we must see what scientific support Pannenberg can provide for it. On the basis of the foregoing description of his methodology, we should expect him to begin with the relevant texts, explicating them first by relating the text to its larger context, and then to confirm the larger context by showing that it also accounts for aspects of current experience. He must also show that the (authentic) resurrection narratives show no positive analogy to legend, myth, or the like.

Pannenberg notes that in the earliest texts the tradition of Jesus' appearances to his followers is separate from the empty-tomb tradition; as time goes by they become progressively more closely linked. Therefore he treats the two traditions separately. The appearance accounts in the Gospels are so thoroughly imbued with legendary elements that he disregards them and concentrates on Paul's account, especially his report in 1 Corinthians 15:3–8:

> For I delivered to you as of first importance what I also received, that Christ died for our sins in accordance with the scriptures, that he was buried, that he was raised on the third day in accordance with the scriptures, and that he appeared to Cephas, then to the twelve. Then he appeared to more than five hundred brethren at one time, most of whom are still alive, though some have fallen asleep. Then he appeared to James, then to all the apostles. Last of all, as to one untimely born, he appeared also to me.

Here Pannenberg emphasizes the proximity of Paul's reports to the events themselves. Although 1 Corinthians was not written until 56 or 57, Paul's visit to Jerusalem, where he would have met at least some of the witnesses mentioned, would have been only six to eight years after Jesus' death. He concludes: "In view of the age of the formulated traditions used by Paul and of the proximity of Paul to the events, the assumption that appearances of the resurrected Lord were really experienced by a number of members of the primitive Christian community and not perhaps freely invented in the course of later legendary developments has good historical foundation."[16]

Next Pannenberg considers what this experience must have been like. Again, Paul's first-hand account is most reliable, and Paul assumed that his experience was like the others'. Paul's account yields several conclusions: the relation of the appearance to the man Jesus was clear, and the appearance was a *sōma pneumatikon*—a spiritual body, not an earthly body resuscitated. Thus Paul's experience was a most unusual one, describable not in literal terms, but only by means of a metaphor—Jesus had 'awakened' from the 'sleep' of death. Furthermore, Pannenberg notes, the apocalyptic outlook prevalent at that time was a necessary condition for the appearances to be recognized as those of one who had been resurrected from the dead.

Here we see Pannenberg shifting from textual criticism to consider the *horizon of meaning* within which the text must be understood. He claims that we could not make sense of the fact that the disciples interpreted their visionary experiences as appearances of one raised from the dead apart from our knowledge of the apocalyptic thinking of the age. This is also, however, the point at which he makes a move to tie the tradition as so far understood to current experience. He asks whether we can use the apocalyptic tradition not only to account for what the first Christians *thought*, but also to account (to ourselves) for what actually happened. To answer, we must decide whether our own experience warrants an expectation for resurrection similar to that of the apocalyptic tradition. Pannenberg bases his argument for such an expectation on the universal phenomenon of hope. Citing evidence from medical studies to the effect that people without hope

16. Wolfhart Pannenberg, *Jesus—God and Man* (Philadelphia: Westminster Press, 1968), p. 91. Trans. by Lewis L. Wilkins and Duane A. Priebe of *Grundzüge der Christologie* (Gütersloh: Gütersloher Verlagshaus G. Mohn, 1964), p. 87. Hereafter *JGM* and *GC* respectively.

simply do not live, he concludes that most people do have enough hope in the future to make their lives and pursuits seem worthwhile. Then he argues that if there is no expectation (either conscious or suppressed) that death is not the end for the individual, then this universal hope is unjustified. The two available concepts of life beyond death are that of the immortality of the soul and that of resurrection (and transformation) of the entire person. Pannenberg claims that we have now come to see, on the basis of scientific knowledge regarding the dependence of mental functions upon the physical, that the survival of a soul beyond the decay of the body is not intelligible. Thus he concludes that resurrection is the most promising source of hope for life after death.

Pannenberg is not claiming that we must know with any certainty that resurrection awaits us; we must only consider it to be a real possibility such that it provides an interpretive category with which to understand the evidence in the text. If the historian is certain ahead of time that the dead do not rise, then it has already been decided that what happened to Jesus was not resurrection. If, however, resurrection is a viable option for people today in anticipating their own future, and if no better explanation is suggested by the text, then we have warrant for adopting the same explanation of Paul's and the others' postcrucifixion experiences that they themselves adopted spontaneously. Notice that our becoming aware of our own hope for resurrection is the means by which our horizon is "fused" with that of the apocalyptic horizon of the first-century Christians.

Pannenberg also considers the tradition of the empty tomb, concluding that it arose independently from the appearance accounts and thereby supports the conclusion reached so far, not only because it supplies independent evidence but also because the main competing theory to explain the appearances is that they were purely psychological aberrations produced by the disciples' excitement over the empty tomb. All things considered, Pannenberg concludes that Jesus' resurrection is "historically very probable" and therefore ought to be accepted in the absence of any other equally well supported hypothesis.[17]

We can see that in his investigation of the resurrection Pannenberg is faithful to his own methodological prescriptions. Jesus' resurrection in turn is the foundation of Pannenberg's dogmatic theology.

17. JGM, p. 105; GC, p. 103.

For example, it is the resurrection that vindicates Jesus' otherwise blasphemous claims that he is entitled to rewrite the law and that the last judgment will be made on the basis of one's reaction to him. In other words, the resurrection is the key to Christology, and Christology is the key to the rest of theology—most important, to the Christian doctrine of God.

If we were to represent Pannenberg's view of theology in spatial terms we might describe it this way: Theology begins with two sets of concentric circles. One set involves the texts of the Christian Scriptures. In Pannenberg's theology the texts concerning Jesus' resurrection are at the very center, surrounded by circles representing the increasingly broad contexts required for their full understanding—most immediately, the rest of the New Testament. The apocalyptic worldview of the first century would in turn form an important context for understanding the New Testament.

Beside this first series of concentric circles is another representing our own experience and the ever-broadening contexts within which we attempt to interpret it. At some point moving outward we reach a circle that encompasses both the texts and our own experience. In the example of the resurrection, this circle would represent an understanding of human life as destined for a transformed future beyond death. This circle represents the immediate context for understanding the apocalyptic worldview mentioned above. The Christian view of God, based upon Jesus' resurrection, is finally represented by the outermost circle, intended to encompass and give meaning to all experience and to all other contexts of meaning.

In propositional form we might summarize Pannenberg's view of theology as follows:

(1) 'God' is defined as the all-determining reality.

(2) Religious traditions are interpretations of special events through the course of history in which the all-determining reality is believed to have made itself known.

(3) Philosophy of history raises (but cannot by itself answer) the question of the source of the unity of history. To raise this question is to raise implicitly and formally the question of God as the all-determining reality.

(4) The theologian, working as a scientist, examines religious traditions as hypotheses about the nature of the all-determining reality, judging them on the basis of how well

(coherently, parsimoniously, accurately) they account for
the totality of historical experience up to the present.

(5) The theologian working within a tradition seeks to understand that tradition "out of itself" and in relation to its
historical context, proposing reinterpretations that are at
once faithful to the original tradition yet responsive to the
pressures of new experience.

(6) The Christian tradition asserts that the God of Abraham,
Moses, and Jesus is the ultimate determinant of the entire
course of history and will be clearly shown to be so on the
last day.

(7) This claim, however, has already been confirmed by the
exceedingly important historical fact that Jesus, the putative spokesman for this God, was raised from the dead.
This resurrection both confirms Jesus' understanding of
God and allows us to know in advance how history will
end on the last day.   -

(8) Insofar as Jesus' resurrection is supportable by scientific
(historical) investigation, the whole of Christian theology
as understood in relation to this event is given scientific
status. It is a hypothesis about the meaning of the totality
of history—already confirmed to some extent but open
to further confirmation as we continue to investigate
whether and how it can shed light upon all other knowledge.

4. The Hume-Pannenberg Debate

Pannenberg's system is impressive in scope and coherence. Furthermore, his assessment of what is needed if theology's cognitive
claims are to be rationally supportable today is, in my view, right on
the mark. The critical question to be addressed in this section, however, is whether or not Pannenberg has met Hume on his own terms
and has provided the kind of support for theism about which Hume
was so skeptical. The crucial issue will be whether Pannenberg's
system provides an adequate theory of scientific method.

4.1 Hume

In order to evaluate Pannenberg's achievement against Hume's
agnostic arguments, we must see Hume's work in its broader con-

text. Donald Livingston's book, *Hume's Philosophy of Common Life*, provides a valuable insight into Hume's philosophical work by placing it against the background of his interest in history, and it is at the level of philosophy of history that the points of agreement and disagreement between Pannenberg and Hume can best be seen.[18] According to Livingston, one of Hume's major goals was to demolish the *providential* view of history, a view firmly established in Hume's time both in popular consciousness and in the metaphysical framework within which the scientists of the age worked. Joseph Priestley, for example, believed that history promised to provide the most important source of insight into divine activity and plan. As we learn more of history, we see more and more clearly the perfections and providence of God, and that the "grand catastrophe" is growing nearer.[19] David Hartley believed that biblical prophecy could be scientifically established and used to understand the historical significance of his own age.[20]

Livingston points out that for Hume the moral world is woven together by the narrative imagination, which registers the passions and thoughts we have about our temporal involvement with natural objects and with one another. "It is a world of narrative associations, that is, of stories having varying degrees of significance and generality." The question then arises for Hume whether these stories might form a system of stories, whether the unities of action that constitute them are, in turn, part of some larger story. "The providential and prophetic view of history is just such a story, and Hume most explicitly rejected it."[21]

Hume's arguments are found in *An Enquiry concerning Human Understanding* (1748).[22] In section x, entitled "Of Miracles," he extended his argument against belief in miracles to include prophecy. If prophecy exceeds our ordinary human powers to make predictions—that is, if it is genuine knowledge of the future—then it must be accounted a miracle, and all the objections (summarized in Chap-

18. Chicago: University of Chicago Press, 1984.

19. Joseph Priestley, *Lectures on History and General Policy* (Dublin, 1778), pp. 452–53.

20. David Hartley, *Observations on Man, His Frame, His Duty, and His Expectations*, 2 vols. (London, 1769).

21. Livingston, *Hume's Philosophy*, p. 149.

22. References are to L. A. Selby-Bigge's edition (Oxford: Clarendon Press, 1902). Hereafter *Human Understanding*.

ter 1 above) apply. The following section, entitled "Of a Particular Providence and of a Future State," contains Hume's objections to the view that we can know by means of reason, based on our experience of the moral order in history, that God is perfectly benevolent and that he therefore has a plan for the perfection of the human condition or future punishment of the wicked. Hume argued that in reasoning from effects to causes we are justified in attributing to the cause only those properties needed to explain the observed effect. Thus God may be justifiably attributed a benevolence proportionate to the amount of justice or progress actually observed in history, but there is no justification for inferring from what we have seen that God is *more* just or benevolent and therefore has a "more finished scheme or plan, which will receive its completion in some distant point of space or time."[23] We know no more about God than whatever may be necessary to explain the phenomena we have observed so far, and we certainly know nothing about the future based upon such a conception of God.

(Note that although Hume has granted throughout the main part of his essay that God may be known through his effects in history, and argues as though he seeks only to limit the extent of this knowledge and its usefulness for interpreting history, he ends by expressing doubt that we can know of God at all through these means. The argument so far has traded upon knowledge of causes by means of their effects, but it is really only when two *species* of objects are constantly conjoined that we can argue from the appearance of an object of the second type to the existence of an object of the first type. The universe is one of a kind, however, and so is the hypothesized cause of it, namely God.)

Let us test Livingston's thesis that "one of Hume's main philosophical and historical tasks was to supplant the traditional Christian story line of the creation, fall, and redemption . . . by a new unity of action based along secular and humanistic lines" by seeing how far it can be used as a key to understanding the rest of Hume's work.[24] We shall see that it allows us to understand Hume's agnostic arguments as part of a *positive strategy* to defend his own views of historical methodology and of the origin and justification of moral judgments.

23. Ibid., p. 143.
24. *Hume's Philosophy*, p. 141.

The main thrust of Hume's philosophical work, both his epistemology and his moral philosophy, can be described as a sustained
attempt to demonstrate the limits of reason divorced from the practices and customs of "common life." In both cases knowledge and
judgment must come from experience. Reason serves to clarify concepts derived from experience and can reform the common life, but it
has no Archimedean point apart from experience and custom from
which it can propose theoretical or political systems to replace entirely those that have slowly developed by trial and error. In the
conclusion to Book I of *A Treatise of Human Nature* (1739) he reflects
on the folly of all metaphysical reasoning: The resolution to "adhere
to the understanding," that is, to guide life by reason alone, "if
steadily executed, wou'd be dangerous, and attended with the most
fatal consequences. For I have already shewn, that the understanding, when it acts alone, and according to its most general principles,
entirely subverts itself, and leaves not the lowest degree of evidence
in any proposition, either in philosophy or common life."[25]

Fortunately, however, we are not left to ponder these skeptical
conclusions for long, since nature herself dispels these "clouds."
"Here then I find myself absolutely and necessarily determined to
live, and talk, and act like other people in the common affairs of
life."[26]

The purpose of Books II and III of the *Treatise* and of *An Enquiry
concerning the Principles of Morals* (1751) is to show that philosophical
reasoning concerning human conduct has its sole origin in common
experience. The moral world *is* a set of conventions that have developed over time. The sole task of philosophy is to bring to explicit
awareness and improve the precision of the standards already implicit in these conventions. Hume concludes Book III of the *Treatise*
saying: "And thus the most abstract speculations concerning human
nature, however cold and unentertaining, become subservient to
practical morality; and may render this latter science more correct in its
precepts, and more persuasive in its exhortations."[27]

From this perspective we can see that Hume's argument against

---

25. References are to L. A. Selby-Bigge's edition (Oxford: Clarendon, 1955), pp.
267–68. Hereafter *Treatise*.
26. Ibid., p. 269.
27. Ibid., p. 621.

the prophetic-providential view of history serves the same function with regard to politics as does his argument against speculative metaphysics. A second sort of distortion, comparable to false metaphysics, is the view that political history is guided not by the laws of human nature but by provident design, and that the moral order is founded not upon social convention and pragmatic concerns, but rather upon divine decree sanctioned by threat of eternal punishment. In order to establish his own view of experience and custom as the source of moral and political order, Hume had to dismantle this alternative account; he did so in three steps.

First—and here we are able to see an integral relation between Hume's philosophical and historical interests—he wrote his six-volume work, *The History of England, From the Invasion of Julius Caesar to the Abdication of James the Second, 1688,* from a purely secular point of view in order to show that history can be understood perfectly well without the prophetic-providential mode of interpretation. He set out to account for events entirely by means of the laws of human nature—the natural propensities of the human mind as known from experience. This is in sharp contrast to other historians of his time, Catherine Macaulay, for example, who depicted a political event—the French Revolution—as a case of "the *omnipotent will* more immediately declaring itself in favor of the future *perfection* and happiness of the moral world."[28]

Second, as I have summarized above, Hume argued that we cannot know anything about God from the course of history or from prophecies in Scripture that allows in turn for prediction of the future course of history or of final reward or punishment.

Third, in "Of a Particular Providence and of a Future State" Hume argued that a religious system offering divine reward or punishment is not necessary to uphold the moral-political order. We can see in this life that virtue is attended with its own rewards, namely "more peace of mind . . . and a more favorable reception from the world" and that vice leads to misery.[29] Any argument to the effect that these earthly rewards and punishments provide evidence for a supreme governor of the world and hence for greater (heavenly) rewards is

28. In *Observations on the Reflections of the Right Hon. Edmund Burke, On the Revolution in France, in a Letter to the Right Hon. The Earl of Stanhope* (London, 1790), pp. 20–21.

29. *Human Understanding*, p. 140.

invalid (as a form of the argument from effects to greater-than-necessary cause) and besides is *unnecessary*, because the experience of earthly reward upon which it is based is already a sufficient guide to conduct.

Hume objected to the providential view of history because it distanced people from the common life of habit and custom, built upon the accumulated experience of the race. From an imagined state of perfection outside or beyond history, the whole present order is devalued—it reverses "the whole course of nature as to render this life merely a passage to something further—a porch, which leads to a greater, and vastly different building."[30] Thus the judgments we used to be able to make of particular goods and evils are all relativized. We view all that happens as but a part of the wholly good plan of God and ignore our true source of judgment—common convention.

A related point is this: we know the meaning of an event only when it is in the past, and far enough in the past to enable us to see its consequences. The historian in any narrative composition "traces the series of actions according to their natural order . . . and delineates their most remote consequences."[31] We would therefore like very much to know the future because that alone would allow for accurate interpretion of current events. The prophetic view of history, describing the end of the world and the last judgment, seems to offer this possibility. But whereas Pannenberg believes we can trust such prophecies, Hume argued that neither revelation nor reason provides knowledge of the future.

4.2 Conflict

It will have become obvious by now that Pannenberg and Hume share some views about history but differ on other points—in fact, their differences can better be described as diametric opposition.

A major point of agreement is their view of the centrality of the concept of history. Pannenberg's historicism with its Hegelian roots is quite explicit. Hume's is less explicit but, as we have seen, his

---

30. Ibid., p. 141.

31. This passage is found in a section of chapter III of *Human Understanding* omitted from the Selby-Bigge edition. See C. W. Hendel's edition (New York: Bobbs-Merrill, 1955), p. 34.

historical concerns serve as a central theme capable of unifying his work. They do so most notably in his ethical writings. Regarding the dependence of moral and political theory on history Hume says: "These records of wars, intrigues, factions, and revolutions [of which history informs us], are so many collections of experiments, by which the politician or moral philosopher fixes the principles of his science, in the same manner as the physician or natural philosopher becomes acquainted with the nature of plants, minerals, and other external objects, by the experiments which he forms concerning them."[32]

To emphasize the narrative or historical nature of moral and political concepts in Hume's thought, Livingston has coined the term "past-entailing predicate." Predicates such as 'is a father,' or 'is a queen,' or 'is a constitution' logically cannot be applied to the present unless some past-tense statement is true. For example, 'gambling is illegal' cannot be true unless some statement, such as one about a legislature's action in the past, is also true. In this way, "past-entailing ideas are woven by the narrative imagination and constitute the conceptual framework of the moral world."[33] Social and political legitimacy, in Hume's view, is constituted by these narrative relations holding between past and present existences.

In a sense, though, the whole of knowledge has for Hume a narrative, or past-entailing, basis. He says: "When we entertain, therefore, any suspicion that a philosophical term is employed without any meaning or idea (as is but too frequent), we need but enquire, *from what impression is that supposed idea derived*?"[34] Hume's most famous analysis of a philosophical term is that of 'causation.' The causal connection that we impute to events is a product of a series of *past* experiences. To say that A causes B is to say that in one's past experience As have uniformly been followed by Bs, and that one's mind has become determined to pass from observation of an A to expectation of a B. Without history the world would lack causation.

But if abstract philosophical terms are to be analyzed in terms of past experience, all the more so simple ideas and perceptual properties. To know what 'is a horse' means, refer to your previous sense

32. *Human Understanding*, pp. 83–84.
33. *Hume's Philosophy*, p. 301.
34. *Human Understanding*, p. 22.

impressions of horses. Thus all knowledge is dependent upon nar-
ratively formed cognitive processes. Hume's intent in the *Treatise*
was to explore just these cognitive processes and to account for all
knowledge on the basis of his conclusions.

As I have already mentioned, both Hume and Pannenberg believe
that the significance of an event can be judged only when we know its
consequences, and thus we can have no definitive 'history' of the
present without knowing the future. They draw different conclusions from this supposition, however. Hume claims that if a definitive history depends on knowledge of the future, then our historical
accounts must not be definitive. Hume's biographer E. C. Mossner,
expressing an opinion consistent with Hume's, says that history is
the past interpreted for the present; every age, consequently, requires
its own interpretation of the past and no other will quite do.[35] Pannenberg, on the other hand, argues (transcendentally) that if historical
knowledge depends on knowledge of the future, then we must have
some hypothesis about the future that is in some measure justified by
the fact of our present knowledge—a pre-projected unity of history.

In sharp contrast to Hume's understanding of all concepts as past-
entailing predicates, Pannenberg claims that all knowledge consists
in the use of future-entailing predicates. This is as true of complex
judgments about the entire course of history as it is of the simplest
statement about a material object. The statement that Jesus reveals
the final meaning of history is anticipatory of a particular sort of
climax to the historical process. The statement that this book is red is
anticipatory in that it entails an infinite number of possible future
sensory experiences (for example, that it will appear red under normal lighting, that it will appear black with blue light shown on it).
Disappointment of some of these expectations would require abandonment or modification of the original judgment. The anticipatory
character of experiential knowledge, as mentioned above, is a view
of Karl Popper's that Pannenberg endorses.

We see another point of diametric opposition when we consider
historical method. Hume's historiography is in the tradition of Vico
and the Port-Royal *Logic*: the probability of testimony regarding past
events is to be judged on the basis of knowledge that such things do
(or do not) happen today.

35. *The Life of David Hume* (Edinburgh: Thomas Nelson, 1954), p. 318.

It is universally acknowledged that there is a great uniformity among the actions of men, in all nations and ages, and that human nature remains still the same, in its principles and operations. The same motives always produce the same actions: The same events follow from the same causes. . . . Mankind are so much the same, in all times and places, that history informs us of nothing new or strange in this particular. Its chief use is only to discover the constant and universal principles of human nature, by showing men in all varieties of circumstances and situations, and furnishing us with materials from which we may form our observations and become acquainted with the regular springs of human action and behaviour.[36]

It is upon just this methodological assumption that Hume based his argument against using miracles or prophecy to validate revelation.

Pannenberg recognizes that this is exactly the assumption that destroys the rational credibility of Jesus' resurrection. He chooses, however, to deny the necessity of the methodological presupposition rather than to give up historical knowledge of the resurrection. This methodological disagreement is the foundation of the opposing conclusions reached by the two authors. Hume's nomological approach to historiography defines historical knowledge as knowledge of regularities—the goal of historical investigation is "to discover the constant and universal principles of human nature," and, furthermore, purported events that violate these regularities could never be adequately confirmed. Pannenberg's "systems" approach allows for knowledge of unique events, but has the consequence of making a final interpretation of events dependent upon knowledge of the whole course of history.

This last requirement motivates Pannenberg to develop his own version of the prophetic-providential view of history. In contrast, while Hume recognized the value of the perspective of distance, he set out to show that knowledge of the regularities of human behavior *is* adequate for interpreting historical events (although not for interpreting the present). Hume rejected the providential view of history in favor of an anthropocentric understanding because he saw the former to have unacceptable consequences, namely, the severing of human evaluations from their true source in the common life. Pannenberg rejects the anthropocentric approach to history because he

36. *Human Understanding*, p. 83.

sees it to have unacceptable consequences, namely, the fragmentation and relativization of history.

It is not entirely clear whether Hume believes in the existence of a creator God.[37] He seems to think that some sort of creator is possible, even probable, but that we can know next to nothing about such a being's qualities—an intelligence as much like ours as ours is like the rotting of a turnip.[38] That Hume would consent to call this the "all-determining reality" is far from clear.

Whereas for Pannenberg the unity of history is a necessary precondition for historical study, Hume thought that any perceived unity is a result *only* of our perception and that we have no ground for assuming it has any other cause. Any knowledge of the unity of history would have to be purely a posteriori and could not be assumed.

In conclusion, we see that Hume and Pannenberg both offer comprehensive systems of thought in which content and method are intimately related. For Pannenberg, the resurrection is a central feature of the content of his worldview because of what it tells us about the end of history, which is important because of his historical methodology, which in turn entails his option in favor of future-entailing predicates, and which is also required by his understanding of the nature of revelation. Because it is just at the point of method that the source of the divergence between Pannenberg and Hume arises, the issues between them cannot be resolved in any straightforward manner; there is no *neutral method* by which to do so.

4.3 Defeat for Pannenberg

We must consider now whether Pannenberg has advanced the argument for theism *beyond* Hume's agnostic position. I shall present two reasons for thinking he has not:

4.3.1 Hume and Pannenberg: Incommensurable Pannenberg has surely presented an alternative, but he must show it to be rationally

37. For a recent discussion, see J. C. A. Gaskin, *Hume's Philosophy of Religion* (London: Macmillan, 1978).

38. *Dialogues concerning Natural Religion* (npp, 1779). Reprinted in *David Hume on Religion*, ed. Richard Wollheim (Cleveland: World, 1963), pp. 193–94.

preferable to Hume's. The most charitable way to address the question whether Pannenberg's system is an advance over Hume's would
be to use Pannenberg's own methodology and related standards of
rationality: what, *in Pannenberg's terms*, would constitute such an
advance? The answer is that if Pannenberg's system could be shown
to be the more comprehensive of the two—if Hume's system could
be encompassed within and interpreted by Pannenberg's—then the
latter would demonstrate its rational superiority. That is, Pannenberg's system must provide the best explanation of all that is. Hume's
system is a part of historical reality; therefore Pannenberg must be
able to explain it. Explanation for Pannenberg means broadening
one's horizon to encompass alien thought-forms. Therefore, by his
own account, to show the acceptability of his system, once confronted with Hume's, he must be able to incorporate what he judges
to be true of Hume's thought into his own—a tall order.

I suggest that, in fact, the alien Humean system cannot be incorporated without destroying Pannenberg's theology. This is not because
they assign different truth values to some statements—for one can
often 'explain' the opponent's errors in one's own terms. The trouble
is that the acceptance of Hume's understanding of the world—with
the historical method it entails and couched in its past-entailing
concepts—*suspends Pannenberg's fact-constituting principles*. Thus, in
Paul Feyerabend's terms, the two systems are *incommensurable*.

Feyerabend developed his incommensurability thesis to account
for differences between comprehensive theories in science, such as
Newtonian and Einsteinian physics, but the term is applicable to
nonscientific systems as well. His notion of incommensurability is
often confused with Kuhn's, so it is worthwhile to consider it in
some detail:

> We have a point of view (theory, framework, cosmos, mode of
> representation) whose elements (concepts, 'facts', pictures) are built
> up in accordance with certain principles of construction. The princi-
> ples involve something like a 'closure': there are things that cannot be
> said, or 'discovered', without violating the principles (which does *not*
> mean contradicting them). Say the things, make the discovery, and
> the principles are suspended. Now take those constructive principles
> that underlie every element of the cosmos (of the theory), every fact
> (every concept). Let us call such principles *universal principles* of the
> theory in question. Suspending universal principles means suspending

all facts and all concepts. Finally, let us call a discovery, or a statement,
or an attitude *incommensurable* with the cosmos (the theory, the framework) if it suspends some of its universal principles.[39]

What Feyerabend means by this can best be conveyed by an example:
he examines the change in art from the schematic archaic style to
realism. His goal is to show that there are no neutral objects that can
be represented in any style whatever; instead members of the societies that produce archaic art (the 'frontal' art of the Egyptians,
early Greek art, primitive art, and the like) actually see and conceive
of the world differently than do members of, say, classical Greek
society. From the features of archaic art, together with an examination of literary style, sentence construction, grammar, and other
clues, he concludes that there are grounds for attributing to the
ancient Greeks an underlying worldview radically different from,
and incommensurable with, that of classic Greece.

The most striking feature of archaic art is that the parts (parts of
bodies, 'parts' of the action, personal characteristics of the actors) are
simply *added* together. For example, in a scene showing a kid being
eaten by a lion, "the lion looks ferocious, the kid looks peaceful, and
the act of swallowing is simply *tacked on* to the presentation of what a
lion *is* and what a kid *is*." This is a *paratactic aggregate*; the elements are
all given equal importance; the only relation between them is sequen-
tial, and therefore no part is presented as determining the others.
Thus "the picture *reads*: ferocious lion, peaceful kid, swallowing of
kid by lion."[40]

Feyerabend argues that this style of art represents a thorough-
going cosmology: the world is actually seen as a paratactic aggregate
of parts, not a system with some underlying source of order and
unity that subordinates some features, parts, or actions to others.
Closely related is the concept of knowledge, thought of not in terms
of depth of understanding, but only in terms of more or less quantity. Quantity of knowledge comes simply from exposure to many
parts of the world; there is no essence to be grasped behind appearances. In fact, there are no *appearances* of things in this world.
There are simply the things themselves.[41]

39. Paul Feyerabend, *Against Method* (London: New Left Books, 1975), p. 269.
40. Ibid., pp. 233–34.
41. Ibid., pp. 260–61.

It is exactly at this point that the contrast between archaic and classical Greek thought can best be apprehended. The new cosmology, arising between the seventh and fifth centuries B.C., distinguished between much knowing and true knowledge. True knowledge was not of the chaotic world of appearances, but rather of the alleged true world that 'lies behind' appearances and is dimly reflected in them.

Each of these two cosmological schemes is made up of different elements. Objects in the earlier one are aggregates of equi-important, perceptible parts; objects in the second are imperceptible essences underlying a multitude of deceptive appearances. Furthermore, the elements of one cosmos cannot simply be added to those of the other. There is a structural change from archaic to classic cosmology that prevents the coexistence of aggregate objects and essence-appearance objects. This change affects many other aspects of life. For example, the artist no longer draws a catalog of persons and events on the surface of the paper, but instead seeks to manipulate appearances (illusions) that will suggest to the viewer a three-dimensional scene.

I suggest that Hume's and Pannenberg's systems are incommensurable in the sense just specified, because of differences in "fact-constituting principles." In fact, the relation between Hume's and Pannenberg's theories of knowledge is rather similar to that between archaic and classical Greek theories. For Hume, facts are constituted by the application of concepts that develop through repeated (past) impressions (of sensation and reflection). Knowledge is atomic, built up by conjunction of and reflection upon individual ideas. It is the atomic nature of knowledge that makes Hume's theory comparable to archaic thought. For Pannenberg, on the other hand, facts are meanings constituted by interpretations of experiences that always depend essentially upon anticipations of the future. Fact-constituting interpretation involves placing the thing to be understood in a context within which its meaning becomes apparent, and these subcontexts depend in turn upon a hypothesis about the meaning of the totality of history.

We can substitute the word 'meaning' for 'essences' and 'appearances' in Feyerabend's contrast between archaic and classical thought, and produce an equally apt contrast between Hume and Pannenberg: "[For Hume] . . . knowledge comes simply from exposure to many parts of the world; there is no [meaning] to be grasped behind appearances. In fact, there are no [meanings] of things in this world. There are only the things themselves.''

Therefore, although Pannenberg can ''critically reinterpret'' and incorporate much of the content of Hume's thought, he cannot incorporate the *epistemological theory* upon which it is based without either changing Hume's views into something essentially non-Humean or else destroying the epistemological basis of his own system. Consider the parallels with styles of art: a trompe-l'oeil painting can incorporate representations of the *objects* of another (primitive) painting by doing them over in its own style ('critical reinterpretation') but it cannot incorporate the *style* of primitive art without giving up its own style and becoming primitive art. (Try to think how a trompe-l'oeil painting could represent a primitive painting's *representation* of a three-dimensional object.) Pannenberg cannot incorporate Hume's fact-constituting style except by giving up his own.

Thus Pannenberg has indeed offered an alternative to Hume but, if I am right about the incommensurability of the two systems, he cannot offer any reason *consistent with his own system* why we should choose his in preference to Hume's, because such superiority (by his own account) depends on the possibility of incorporating Hume's system into his own.[42] They are both comprehensive, well-argued systems that give meaning to the entire historical process.[43]

If Feyerabend is right, we could have predicted that Pannenberg's methodology would run aground somewhere—two incommensurable horizons cannot be merged by broadening one to incorporate the other, as Pannenberg's methodology requires. But we could not know ahead of time that Hume's system was one that would destroy Pannenberg's. That Pannenberg himself is not sufficiently aware of the possibility of incommensurable frameworks is indicated by the

42. Of course, Hume's system cannot incorporate Pannenberg's either, but the success of Hume's methodology is not predicated on such all-inclusiveness, so the Hume-Pannenberg stand-off remains.

43. Is there any other criterion consistent with Pannenberg's theory of knowledge by which to judge his system superior to Hume's? Perhaps Pannenberg would say that his whole is more meaningful than Hume's because it satisfies our need to know the end of history in order to find meaning in life. But Hume could reply that we only think we need to know the end of history; if we did know, then the real meaning of the present life would be annulled.

fact that in his discussion of scientific explanation he commends, as a refinement of Toulmin's theory, that proposed by Erhard Scheibe, which requires of any new theory put forth as an explanation that it (among other things) account for how the previous theory fits into the new one. This modification of Toulmin's theory "makes it even clearer that explanations of this sort also function by fitting parts into a semantic whole . . . [such that it] shows the previously contradictory elements as reconcilable in the semantic web of a new systematic."[44]

4.3.2 Pannenberg's Methodology: Unworkable   We could also have been at least suspicious beforehand that Pannenberg's methodology would prove inadequate had we considered the philosophy of science to which it is most closely tied and noted how that approach has fared in recent philosophy of science discussions. Toulmin is not guilty, as Scheibe is, of failing to recognize the possibility that a new theory in science may not be reconcilable with its predecessor in such a way that the latter is included in or interpreted by the former. In fact, Toulmin was one of the first of what may be called the historicist school in contemporary philosophy of science (a school that includes Feyerabend, Thomas Kuhn, Imre Lakatos, and others), one of whose major contributions has been to call attention to this very issue. Nonetheless, Toulmin's contribution to philosophy of science in the volume Pannenberg cites, *Foresight and Understanding*, is largely descriptive and critical.

When Toulmin turns in the last chapter to consider the problem of determining which of two rival explanations (or ideals of natural order upon which they are dependent) yields greater understanding, he turns to the language of evolution. "In the evolution of scientific ideas, as in the evolution of species, change results from the selective perpetuation of variants." This leads him to phrase the question about the merits of scientific theories in terms of their survival value. Biological species survive and evolve not by meeting any single evolutionary demand but because they alone, from the available

44. *TPS*, pp. 151–52; *WT*, pp. 152–54. Pannenberg has said (in private conversation, April 10, 1987) that he can incorporate a critical reinterpretation of Hume, that is, everything that he judges to be *true* in Hume's system, but I maintain that the crucial question is whether Hume's and Pannenberg's fact-constituting principles are compatible, and such principles are not the sort of thing that can be true or false.

variants of earlier forms, have successfully met the multiple demands of the environment.[45] But in these terms no abstract *theory* of confirmation can be devised that will serve as a basis for choosing between theories.

In sum, just as evolutionary theory provides no criterion for determining which species ought to survive—we can judge survival *value* only on the basis of actual survival—so in science, once theory choice is assimilated to survival of the fittest, Toulmin recognizes that he can offer no general guidance for theory choice. An evolutionary approach to philosophy of science may be helpful as a description of the progress of science after the fact, but provides no prescriptive methodological theory. Consequently, the theologian who turns to Toulmin's philosophy of science may be aided in describing the evolution of theological theories, but should not expect to find help in settling tough questions of choice between competing explanations such as Pannenberg's and Hume's opposing theories of history.

Finally, a few words about Pannenberg's theory of revelation. He opted for the view of revelation as the indirect manifestation of God through the whole of history in order to have an 'objective' starting point for theology—one not requiring the eyes of faith. But it seems that his system has failed just at that point. We have seen that an entirely different interpretation of history is possible, for Hume and his followers see the course of history devoid of Providence, where Pannenberg sees it as a series of acts of God.[46] Pannenberg claimed that what is needed to settle disputes such as that between Hume and himself is for the historians to get straight about the facts: was Jesus raised from the dead or not? But getting straight about this fact led to disputes about the proper method for historians to use, and the choice of method seemed to depend in the end on abstruse views about the meaning of history and whether or not one could account for it without assuming a transcendent ground of the unity of history—which brings us full circle.

So my second criticism of Pannenberg's methodology is that it is simply unworkable. When detection of the action of God in the

45. *Foresight and Understanding*, pp. 110, 111.

46. Christians must admit that the Humean, naturalistic account of history (and nature) has won the day in academia. So the competing interpretation is not a bare possibility, but rather the interpretation of choice for the majority of historians (and scientists).

world is made to depend upon judgments of vast scope, such as that concerning which of two entire worldviews does a better job of giving meaning to the totality of experience, we must simply throw up our hands in despair. We may have opinions on such matters, but we have no criteria for settling disputes of this sort when they arise.

In Chapter 5 we shall examine Christian 'epistemic practices' designed to yield intersubjective judgments regarding putative divine actions in particular cases. Pannenberg claims that we do not need a prophet to interpret history. I shall claim that we do, and that testing the prophets is a major epistemic task for the Christian community, yet not a task immune from objective test and rational criticism.

In Chapter 6 I shall return to Pannenberg's theology, indicating how the methodology developed in the following pages can be used to reconstruct the theological content of Pannenberg's system in such a way that its scientific character (in the Anglo-American sense of 'science') becomes clear.

[3]

Probable Reasoning Come of Age—Philosophy of Science

Imre Lakatos, who died suddenly . . . in February 1974, was a fascinating person, an outstanding thinker and the best philosopher of science of our strange and uncomfortable century. He was a rationalist, for he thought that man had the duty of using reason in his private affairs as well as in any inquiry concerning the relation between himself, nature, and his fellow man. . . . He had realistic ideas about this capability of reason for he emphasised that . . . if reason is to have a point of attack in this world . . . it must be both sly and sophisticated. In the sciences, for example, reason must not be more primitive than the theories which it is supposed to evaluate.

—PAUL FEYERABEND, "Imre Lakatos"

If theological knowledge claims are to be evaluated against the standards of probable reasoning, it is important to begin with the best available account of such standards. Only in this generation has a theory of scientific reasoning appeared that is sophisticated enough to make a meaningful assessment of theology's scientific status. The theory appears in the writings of Imre Lakatos.

To appreciate the value of Lakatos's work one must see it against its historical background. Ian Hacking suggests that it is important to take into account Lakatos's continental European (Hungarian) origins as well as the British audience for which he wrote his philosophical works.[1] Lakatos brought with him to the London School of

1. In "Imre Lakatos's Philosophy of Science," *British Journal for the Philosophy of Science* 30 (1979), 381–402. For biographical information on Lakatos, see John Worrall, "Imre Lakatos (1922–1974): Philosopher of Mathematics and Philosopher of

Economics a Hegelian conception of the development of modern philosophy. His early education had convinced him that Kant had shown the representational theory of truth, to which English-language philosophy was still by and large attached, to be untenable. Consequently, Lakatos set out to provide a theory of scientific rationality that could account for the growth of knowledge and maintain the English philosophers' values of objectivity and critical discussion without recourse to a representational theory of truth.

To refer to the tension between British empiricism and the historicist philosophical tradition of continental Europe is to paint the background to Lakatos's thought in the broadest possible strokes. To see the fine detail we shall have to consider the work of Lakatos's predecessors and contemporaries in philosophy of science.

1. Historical Background

Histories of the philosophy of science often begin with the logical positivists, members of the Vienna Circle and others, who in the 1920s revolted against Kantian and Hegelian idealism. They had two major goals: to distinguish between science and metaphysics (the problem of demarcation), and to reconstruct all (scientific) knowledge from experience (the task later to be called the problem of the confirmation of theories). These two problems, somewhat differently understood through the history of philosophy of science since then, have provided the main structure for work in the field. The verificationist theory of meaning was the logical positivists' proposal both for a criterion of demarcation and for a theory of confirmation. With regard to demarcation, the *meaning* of a statement was to be given by stating the operations that would verify it. Statements whose meaning could be specified in this manner were counted scientific; all else was not only not science—it was meaningless.

Most of the logical positivists answered the question of confirmation by declaring that scientific theories were to be verified on the

Science," in *Essays in Memory of Imre Lakatos: Boston Studies in the Philosophy of Science* 39 (1976), ed. R. S. Cohen et al., pp. 1–8; and Paul Feyerabend, "Imre Lakatos," *British Journal for the Philosophy of Science* 26 (1975), 1–18.

basis of sense data.[2] Now, sense data are an unhandy starting point: they occur only once; they are private; our language is not apt for referring to them—in fact, it takes special (philosophical) training to learn to describe one's experience in terms of brown canoid appearances (say) rather than as seeing brown dogs. But the foundationalist doctrine requiring a *certain* basis for knowledge made the incorrigible reports of appearances seem preferable to corrigible reports about real things—it might not be a dog after all. Rudolf Carnap, one of the most notable members of the Vienna Circle, set out upon a program to construct (as in mathematics) the physical world out of "protocol sentences," each a report of an experienced property with a set of space-time coordinates.[3] Carnap and others, however, soon found the difficulties of this task and other similar projects involving induction to be insuperable. A further difficulty arose when they noted that the verificationist principle itself was not verifiable and thus, if they were to avoid inconsistency, had to be qualified or abandoned.

Given the difficulties faced by the logical positivists, one can imagine the excitement of young Karl Popper, an associate of the Vienna Circle, when he found that he could circumvent both the problem of induction and Carnap's admitted failure at construction with his own new theory of demarcation—his new theory of what makes science scientific. According to Popper, science is characterized by the fact that its theories are falsifiable.[4] He claimed that the method of science is to propose bold conjectures and then to try repeatedly to show them false. This solves the problem of induction: whereas no finite number of observations can ever prove a universal generalization (the logical form of most scientific theories), it takes only one inconsistent observation to falsify it.

This simple logical point, Popper recognized, needed sophisticated qualifications in order to be applicable to science. He was well aware that a single negative test of a theory need not be counted as falsifying it, because the test results could always be questioned.

2. A few, such as Otto Neurath and Rudolf Carnap in his later writings, were physicalists.

3. Rudolf Carnap, *Der logische Aufbau der Welt* (Berlin, 1928).

4. Karl Popper, *Logik der Forschung* (Vienna, 1935), trans. Popper et al. as *The Logic of Scientific Discovery*. Quotations below are from the Harper Torchbook edition, New York, 1965.

Also, no theory has observable consequences standing alone, but only with the addition of initial conditions, auxiliary hypotheses regarding instrumentation, and so on, and a *ceteris paribus* clause to the effect that nothing is interfering with the experiment. Therefore theories are never falsified immediately, because any of these other premises may be called into question instead and will then have to be subjected to falsification tests of their own. Popper worked out a rather elaborate scheme for assessing degrees of falsifiability in order to show that there is a positive relationship between falsifiability and the amount of information a theory provides about the world. His methodology called for the acceptance of the theory that is the most falsifiable yet not in fact falsified.

Popper's methodology represented a decisive shift in approach to both the central problems addressed by the logical positivists. With respect to demarcation, it was not verifiability but rather the ability of a theory to be falsified (or the scientist's ability and willingness to state in advance what will count as falsifying it) that was to be the hallmark of science. Popper's theory of demarcation was not a theory of meaning. Therefore what is not scientific (his examples were psychoanalysis and the Marxist theory of history) is not necessarily nonsense, as it was for the logical positivists. The methodology itself is not supposed to be falsifiable, but rather, it has the status of a set of rules for 'playing the game' of science. A second important change was in the data: Popper's basic statements were reports of repeatable experiments or observations rather than the philosophers' sense data. Such reports are not incorrigible; if called into question they can always be tested by attempts to falsify further observable consequences drawn from them. Here we see the beginning of the end of the logical positivists' foundationalism in that science is no longer seen to rest upon an *indubitable* foundation. Popper still saw the structure of science to be that of an edifice built upon observation statements, but he used the metaphor of pilings driven into a swamp:

> The empirical basis of objective science has thus nothing 'absolute' about it. Science does not rest upon solid bedrock. The bold structure of its theories rises, as it were, above a swamp. It is like a building erected on piles. The piles are driven down from above into the swamp, but not down to any natural or 'given' base; and if we stop driving the piles deeper, it is not because we have reached firm

ground. We simply stop when we are satisfied that the piles are firm enough to carry the structure, at least for the time being.[5]

Whereas Popper answered the problem of confirmation by turning it on its head and requiring instead the absence of falsification, other thinkers of his generation—other neopositivists—sought to develop straightforward theories of confirmation. Carl Hempel is perhaps the best example here. As did Popper, Hempel took the logical relation between theory and observation to be deductive, *from* theory *to* data. But he concentrated on positive rather than negative outcomes. According to Hempel, scientific reasoning proceeds in the following manner: One begins with statements describing observations. Then one invents a hypothesis that (along with the addition of various auxiliary hypotheses, statements of initial conditions, and the like) deductively implies the observation statements. The hypothesis (a law or theory) is then tested by the deduction of further observable consequences from it, which, if borne out by experiment or observation, are taken to confirm the hypothesis. Likewise, when a consequence is not borne out the theory is disconfirmed. Hempel named this the hypothetico–deductive model of scientific method. He further linked this form of reasoning with explanation: a fact is explained when we have found the best hypothesis from which it can be deduced, making confirmation of hypotheses and explanation of events isomorphic. Thus scientific theories were to be understood as hypotheses proposed in order to explain observational data and confirmed by having further testable conclusions deduced from them.[6]

Popper, Hempel, and other neopositivist philosophers of science presented a tidy and convincing account of the "logic of science." In their works science appeared a paragon of rationality, subject to cumulative growth and objective testing of all of its assertions. The

5. Ibid., p. 111.

6. See his *Aspects of Scientific Explanation* (New York: Free Press, 1965); and *Philosophy of Natural Science* (Englewood Cliffs, N.J.: Prentice-Hall, 1966). I shall employ a weakened version of Hempel's concept of explanation here. A fact is explained by a hypothesis (law or theory) if that fact follows from the hypothesis (with the addition of auxiliary assumptions). The logical relations need not be strictly deductive, nor does the hypothesis have to be a universal generalization. Lakatos's criterion of empirical progress, then, is to be used to arbitrate between competing explanations. See below, chap. 6, sec. 1.3.6, for a more detailed account.

incorrigibility of sense data and the certitude of strict verification sought by the logical positivists had been given up, to be sure, but given a careful process of conjecture and testing, Truth could be expected not to elude science's grasp for long. This picture of science set the stage for the critical work of historians and philosophers of science such as Stephen Toulmin, Paul Feyerabend, and Thomas Kuhn.

Toulmin was one of many to criticize Hempel's deductive-nomological view of explanation: deducibility from laws of the phenomenon to be explained is too narrow a concept of explanation. Toulmin claims that an event is explained whenever it is placed within a context that makes it intelligible.[7]

Thomas Kuhn's *Structure of Scientific Revolutions* is one of the best-known books in philosophy of science today.[8] Kuhn's critics often accuse him of irrationalism, but his work is better seen as a statement to the effect that (given the assumption that science as practiced *is* in the main rational) the actual history of science does not support the neopositivist view of scientific rationality. In a way Kuhn is like Hume. Hume came upon the scene when Cartesian rationalism and Baconian inductivism were the standard accounts of rationality, and pointed out that ordinary thinking simply does not operate according to these theories, and could not. One could have concluded from this fact either that no one is rational or that rationality is something different from what the theories say it is. Similarly, Kuhn came upon the scene when neopositivist philosophy of science provided the standard account of rationality, and pointed out that science simply does not operate that way.

Kuhn's major innovation was to show the dependence of theory choice in science on factors other than observation and logic. He described the history of science as a succession of paradigms (standard examples of problem solutions in a field) that carry with them laws, theories, applications, and instrumentation. Those who share the view of science associated with a particular paradigm share the same rules and standards for scientific practice, but when paradigms change so do these standards. Kuhn's contribution to the theory of

7. Stephen Toulmin, *Foresight and Understanding* (New York: Harper, 1961), p. 81.

8. 2d ed. (Chicago: University of Chicago Press, 1970).

confirmation was to show that it is the paradigm as a whole—that is, exemplar plus associated theories, standards, and so on (later termed a "disciplinary matrix")—that scientists accept, and that the acceptance of a paradigm involves at the same time acceptance of a certain domain of data that are interpreted in terms of the paradigm's view of the world. Thus there is no pool of independent data that allows scientists to arbitrate between competing paradigms. They are in fact chosen for any number of reasons: for instance, one paradigm may solve the problem that led its competitor into crisis; or it may appear preferable on the basis of one or more values such as simplicity or accuracy of empirical fit. None of these factors singly or in combination, however, will uniquely determine the 'true' paradigm.

Once a paradigm is accepted, its basic laws and theories are not subjected to testing, as Popper and Hempel claimed, but rather they are *assumed* and used for solving the many problems that a paradigm encounters. Thus a comprehensive view of the subject matter is accepted all at once for a variety of reasons and is maintained 'dogmatically' until its problem-solving ability is seen to be outweighed by anomalies, and a new paradigm appears to take its place.

Kuhn's philosophy represents the total replacement of foundationalism with a holist view of science. Here the data are not merely subject to falsification, as in Popper's theory, but are partially determined by the paradigm they 'support.' At this point the metaphor of a theoretical structure *supported* by facts is no longer as appropriate as it was for the neopositivists with their 'layer cake' model of science. The paradigm, including its data, must be accepted or rejected as a whole. Criteria for acceptance involve global features of the system: simplicity, empirical fit, and the like.

Kuhn's approach to the problem of demarcation differed from that of both the logical positivists and Popper. He offered a criterion for distinguishing between mature and immature science. Mature science is that which has reached the stage where one paradigm at a time dominates research. Immature science is preparadigmatic research where a number of schools compete for the field and as a consequence more time is taken in debating presuppositions than in pursuing research.

I mentioned in Chapter 2 that Toulmin has difficulty defining progress in science. His methodology, expressed in evolutionary language, is useful for describing changes in science after the fact, but does not allow him to propose a normative methodology. Therefore progress can only be assumed: the theory that survived must have been the one most suited to survive and therefore the best. Kuhn has similar difficulties. Many scientists and philosophers have been captivated by his description of the cyclical changes in science: paradigm formation, normal scientific research, crisis, revolution, new phase of normal science. But lacking a normative methodology—a theory about what kinds of theories *ought* to be accepted—he is reduced to speaking about those "destined to win" their fight for survival.[9] The only guidance scientists have been able to extract from Kuhn's work is that if one belongs to one of the "immature" disciplines, one must get busy and develop consensus on the field's first paradigm.

We see now that there have been two major waves of positive positions on the problems of demarcation and confirmation: those of the logical positivists and those of the neopositivists. Each of these positions has been effectively criticized: the logical positivists by the neopositivists, and the neopositivists in turn by the historicist philosophers of science such as Kuhn and Toulmin. What is needed next is a new positive position, a normative theory of scientific method that can take into account the historicists' criticisms as well as the information they provide from historical studies of the actual practice of science. This is exactly what the work of Imre Lakatos provides.

2. Imre Lakatos and the Methodology of Scientific
   Research Programs

Lakatos's most influential piece in the philosophy of science is entitled "Falsification and the Methodology of Scientific Research Programmes."[10] Here he made it plain that he intended to continue in the positive (rather than critical) tradition of Popper, while taking into account criticisms of Popper's work by historicist philosophers

9. *The Structure of Scientific Revolutions*, p. 159.
10. In *Criticism and the Growth of Knowledge*, ed. Lakatos and Alan Musgrave (Cambridge: Cambridge University Press, 1970), pp. 91–196. Reprinted in *The Methodology of Scientific Research Programmes: Philosophical Papers*, vol. 1, ed. John Worrall and Gregory Currie (Cambridge: Cambridge University Press, 1978), pp. 8–101. Page references are to the latter.

of science. In Lakatos's view, scientific rationality requires the specification of a criterion for choice between competing "research programs." A research program consists of a set of theories and a body
of data. One theory, the "hard core," is central to the program.
Conjoined to the core is a set of auxiliary hypotheses that together
add enough information to allow the data to be related to the theory.
Examples of types of auxiliary hypotheses are theories of observation
or of instrumentation and lower-level theories that apply the core
theory in different kinds of cases. The auxiliary hypotheses form a
"protective belt" around the hard core because they are to be modified when potentially falsifying data are found. A research program,
then, is a series of complex theories whose core remains the same
while auxiliary hypotheses are successively modified, replaced, or
amplified in order to account for problematic observations.

Lakatos claimed that the history of science is best understood not
in terms of *successive* paradigms, as it is for Kuhn, but rather in terms
of *competing* research programs. Some of these programs Lakatos
described as "progressive" and others as "degenerating." A degenerating research program is one whose core theory is saved by ad hoc
modifications of the protective belt—mere face-saving devices, or
linguistic tricks, as Lakatos called them. We have some sense of what
these expressions mean, but it is difficult to propose criteria by which
to rule out such nonscientific maneuvers. The heart of Lakatos's
methodology is his characterization of the kinds of maneuvers that
are scientifically acceptable. A research program is said to be progressive when the following conditions are met:

(1) each new version of the theory (core theory plus its auxiliaries) preserves the unrefuted content of its predecessor;
(2) each has excess empirical content over its predecessor; that
is, it predicts some novel, hitherto unexpected facts; and
(3) some of these predicted facts are corroborated.

When the first and second conditions are met, a theory is said to be
theoretically progressive. When all three are met it is empirically
progressive as well. From this the contrary follows. A research
program is degenerating when the change from one version to the
next accounts at most for the one anomaly (or set of anomalies) for
which the change was made, but does not allow for the prediction

and discovery of any novel facts. The choice of a theory thus becomes a choice between two or more competing series of theories, and one chooses the more progressive of the programs. Consequently, the choice depends on the programs' relative power to increase scientific knowledge.

Lakatos distinguished, as did Kuhn, between mature and immature science. In mature science, according to Lakatos, a research program includes both a negative and a positive heuristic, which are plans for future development of the program. The negative heuristic is simply the plan (or methodological rule) to avoid falsification of the hard core—to direct the *modus tollens* against the auxiliary hypotheses and make suitable modifications there. This aspect of the heuristic takes account of the fact, emphasized by both Feyerabend and Kuhn, that comprehensive theories need time to develop before they can be judged; progress in science requires a certain amount of "dogmatism." Both mature and immature programs are characterized by a negative heuristic—without the decision to pursue and defend some central theory, there is no program at all.

Mature science differs from immature science in that the development of the auxiliary hypotheses in the former proceeds according to a preconceived plan: the positive heuristic. This is "a partially articulated set of suggestions or hints on how to change, develop the 'refutable variants' of the research-programme, how to modify, sophisticate, the 'refutable' protective belt."[11] In other words, the positive heuristic sets out a program for a chain of ever more complicated models simulating reality. The simpler models are known ahead of time to have numerous counterinstances. But adjusting for these expected anomalies is already anticipated in the plan to elaborate more complex models. In immature science, on the other hand, there may be progressive modifications, but they are random and unplanned. They may be motivated more by external events, such as the need to take account of competing programs' new discoveries.

Lakatos's answers to the two main problems of philosophy of science can now be stated. Regarding demarcation, he claimed that we have *science* whenever there is a series of theories whose empirical content (sometimes) increases as the auxiliary hypotheses are modified to avoid falsification.[12] Furthermore, we have *mature* science

---

11. "Falsification," p. 50.

12. Note that programs need not be progressive all the time in order to maintain

whenever these content-increasing modifications are in accordance with a preconceived plan. As to the problem of confirmation, Lakatos claimed that there is objective reason for choosing one program over another when the former has a more progressive record than its rival—that is, a greater demonstrated ability to anticipate novel facts. It is not necessarily irrational, however, to continue to work on a degenerating program for two reasons: a degenerating program can sometimes be turned around (made progressive again), and competition in the field generally leads to more overall progress.

3. Evaluation of Lakatos's Methodology

It is possible now to offer reasons for believing Lakatos's to be the best account of scientific rationality available. My argument involves three steps: first, an explanation of Lakatos's plan for showing that his methodology is supported empirically by the history of science; second, replies to the most significant criticisms of his work; and finally, evidence that Lakatos's most recent competitor in fact offers no better alternative.

3.1 Scientific Methodology as a Theory of the History of Science

In my brief review of the history of philosophy of science preceding Lakatos, I alluded to the question of the justification of scientific methodology, at least to the extent of noting that the positivists' verification theory was rejected (in part) because it was not itself verifiable; it was rather self-stultifying.

In order to avoid a parallel problem, Popper claimed immunity from falsification for his falsificationist methodology on the ground that he was proposing rules for the practice of science, and rules as such can be neither true nor false. It is not clear that this is an adequate defense, however. Consider the following: A visitor from another culture, unfamiliar with the game of chess, observes the game for

---

their scientific status. In fact, it is not clear what this would mean, since prediction and corroboration of facts are discrete events. A program that has once been progressive and later degenerates will be given a grace period. At some point, however, it will come to be seen as pseudoscientific.

some time and finally announces that she or he has determined the
rules by which the game is played. It is surely possible that the visitor
has been able to discover the actual rules of the game. But it is also
possible that he or she has devised rules different from those by
which the game is played. It would therefore be more accurate to
speak of the visitor's *theory* about the rules of chess, and theories,
unlike rules, can be right or wrong. Now, let us modify the example
somewhat. Suppose that our visitor proposes not a theory about the
rules of chess, which are clear, stable, and directly determinative of
the range of possible moves, but rather, a theory about winning
strategies used by chess masters. In this case we can imagine a wide
variety of theories capable of describing more or less accurately
various episodes in the history of the game.

It was Kuhn's accomplishment to show both that the methodological "rules" proposed by Popper, Hempel, and others did not in fact
account for major plays in the history of science, and that a theory
about scientific method is more like a conjecture regarding good
chess strategy than a list of chess rules.

Lakatos's own contribution to the question is found in a fascinating piece entitled "History of Science and Its Rational Reconstructions" (hereafter, "History").[13] He began the article with the thesis
that the study of scientific methodology involves study of the history
of science in order to determine as closely as possible what goes into
the 'masters'' best strategies. A methodology can be accepted as a
norm—that is, as a set of *pre*scriptions—for further scientific research
only if it is the best supported theory about the actual practice of
science in its finer moments. Thus a methodology plays a double
role: it is a theory intended to account for the successful episodes in
the history of science, and it is therefore descriptive; it is also an
attempt to say how the practice of science ought to go and is thus
prescriptive as well.

Lakatos assumed that science is by and large a rational enterprise,
but this does not entail that every decision scientists have made in
accepting or rejecting theories is based on rational considerations
alone. Therefore Lakatos distinguished between the internal and the

---

13. In *Boston Studies in the Philosophy of Science* 8 (1971), ed. R. C. Buck and R. S.
Cohen, pp. 91–135. Reprinted in *The Methodology of Scientific Research Programmes*,
vol. 1, pp. 102–38. Page references are to the latter.

external history of science. Internal history assumes some theory of scientific rationality (inductivism, falsificationism, Lakatos's own methodology of research programs, or yet another) and explains as much of the history of science as possible on that basis. Philosophers of science refer to this as a rational reconstruction of scientific achievements because the point is to recount the history in a way that makes its underlying rational structure apparent. All aspects of the history of science that cannot be accounted for in terms of the historian's theory of rationality must be explained on the basis of other, nonrational, factors. Thus "external history" is always a necessary complement to internal history. It explains the events for which internal history cannot account, using whatever categories are appropriate: sociological, economic, and so on.

Lakatos began "History" by comparing inductivism, conventionalism, and falsificationism on the basis of their ability to account for the greatest amount of the history of science—that is, on the basis of how much of the history of science each is able to reconstruct as rational.[14] The interesting complication here is that Lakatos's task was to choose among competing *theories* of scientific rationality on the basis of the evidence provided by history, but the very question at issue in the debate over scientific methodologies is the question of what makes for the acceptability of theories, and this applies just as much to theories about the rational structure of the history of science as it does to scientific theories themselves.

Therefore Lakatos first showed that Popper's falsificationism is more suitable than either inductivism or conventionalism for the rational reconstruction (internal history) of science. But he then pointed out that *if* falsificationism is assumed to be the correct account of rational theory choice, then this means that it is an acceptable theory of scientific method only if it is *not itself falsified* by the history of science. So on its own terms it must be rejected because it is, in fact, falsified by many historical incidents.

Next, Lakatos proposed that his own methodology be taken as the hard core of a progressive research program in the history of science. As such, he claimed, it allows for the reconstruction of much more of

14. Inductivism is the Baconian view that science proceeds by means of induction from observations. A conventionalist view of science holds that scientific theories merely provide category systems for the classification of observations.

the history of scientific achievement as rational. He could explain, for instance, why theories faced with anomalies are not in fact rejected out of hand; they are pursued as long as the positive heuristic holds out and new discoveries are being made. Lakatos's papers are full of examples from the history of science that serve as illustrations, but whereas in Popper's work the only function of history is to provide illustrations, in Lakatos's work such examples also provide confirmation for his methodology. Moreover, since Lakatos's "Falsification and the Methodology of Scientific Research Programmes" (hereafter, "Falsification") was published, others have used it to reconstruct various other episodes in the history of science and have thus provided further confirmation for his theory.[15]

It is important to remember, though, that the individual theory is not to be the unit of comparison here; rather it is a series of theories that together form a research program. A curious feature of Lakatos's "Falsification" is that the article begins not with his own views but with Popper's, first highlighting all the features which coincide with his own views. Then he presents his own methodology as but a slight improvement on Popper's. In brief, he claimed to find all the central features of his methodology either implicitly or explicitly in Popper's work except the distinction between mature and immature science, based on the presence or absence of a positive heuristic. The requirement of a positive heuristic is intended to solve the 'anomaly' called the "tacking paradox," which he explains as follows: "According to our definitions, adding to a theory completely disconnected low-level hypotheses may constitute a 'progressive shift'. It is difficult to eliminate such makeshift shifts without demanding that the additional assertions must be connected with the original assertion *more intimately* than by mere conjunction."[16] The more intimate connection is provided by Lakatos's concept of the positive heuristic.

I suspect that Lakatos intended in "Falsification" to locate his own

15. See, for example, Richard Michod, "Positive Heuristics in Evolutionary Biology," *British Journal for the Philosophy of Science* 32 (1981), 1–36; Michael Hallett, "Toward a Theory of Mathematical Research Programmes," *British Journal for the Philosophy of Science* 30 (1979), 1–25 and 135–59; James Cushing, "Models and Methodologies in Current Theoretical High-Energy Physics," *Synthese* 50 (January 1982), 109–24; G. C. Archibald, "Method and Appraisal in Economics," *Philosophy of the Social Sciences* 9 (1979), 304–15.

16. "Falsification," p. 46.

theory of scientific rationality as part of a series of such theories, and therefore as part of a research program in the history-and-methodology of science. It is interesting to note that Popper vigorously rejected Lakatos's reinterpretation of his work, but this was no problem for Lakatos. According to his methodology, "semantic reinterpretations" of a theory are acceptable so long as they lead to an increase in the empirical content of the theory. Lakatos's reinterpretation led to a vast increase in content, because Popper's theory, according to his own interpretation, has no empirical content at all— it is not an empirical theory about science, but a set of rules. With Lakatos's reinterpretation the whole history of science became relevant content for the theory, a fact that constitutes great theoretical progress for Lakatos's version of the program. Each episode in the history of science that fits Lakatos's methodology contributes to empirical confirmation of the program.

So in this article Lakatos considered the best of his competitors' theories of scientific method (falsificationism) and showed that if it is taken as the correct view of the way theories are to relate to their data (as a normative methodology) then, by its own standards, it must be rejected. It is falsified by many episodes in the history of science— episodes in which scientists failed to conform to falsificationist standards. Then he showed that this same *reductio* criticism could not be brought against his own theory because as a theory of the history of science it does meet its own (more lenient) standards for evidential support.

3.2 Criticisms

Responses by philosophers of science to Lakatos's two major papers have fallen roughly into three classifications: (1) papers that use his methodology as a tool for reconstruction of historical episodes and thus provide corroboration;[17] (2) papers comparing Lakatos's work with that of other philosophers of science;[18] and (3) papers

17. See n. 15 above.
18. For example, Janet Kourany, "Towards an Empirically Adequate Theory of Science," *Philosophy of Science* 49 (December 1982), 526–48; M. D. Staffleu, "Theories as Logically Qualified Artefacts," *Philosophia Reformata* 46 and 47 (1981, 1982), 164–89 and 20–40; Gregory Currie, "The Role of Normative Assumptions in Historical Explanation," *Philosophy of Science* 47 (Spring 1980), 456–73.

criticizing various aspects of his methodology. The criticisms are rather varied, and some are confused. The important areas of concern come down to three: (a) Lakatos's methodology depends upon the definition of "novel" facts, and it is not obvious that there is an entirely satisfactory way to explicate this term. (b) Other authors object for various reasons to Lakatos's equation of his (normative) methodology with an (empirical) research program in the history of science and his claim that the success of the latter validates the former. (c) Finally, there is Feyerabend's criticism to the effect that Lakatos's methodology, depending on how it is used, will either harm science by limiting its freedom or else will be merely a "verbal ornament" adorning an unrestricted scientific practice.

Unfortunately Lakatos did not live long enough to clarify his position and respond to critics. In the following pages I attempt to do so in a way consistent with his own hard core.

3.2.1 What Are Novel Facts?   In "Falsification" Lakatos specified that a novel fact is one that is "improbable or even impossible in the light of previous knowledge."[19] Nonetheless, in a paper written after "Falsification" (in conjunction with Eli Zahar) titled "Why Copernicus's Programme Superseded Ptolemy's," Lakatos said:

> I originally defined a prediction as 'novel', 'stunning', or 'dramatic' if it was inconsistent with previous expectations, unchallenged background knowledge and, in particular, if the predicted fact was forbidden by the rival programme. The best novel facts were those which might never have been observed if not for the theory which anticipated it [*sic*]. My favourite examples of such predictions . . . were the return of Halley's comet, the discovery of Neptune, the Einsteinian bending of light rays.[20]

With this understanding of novel facts, however, the Copernican program would not count as progressive until *long after* it had become widely accepted. To account for this anomaly, Lakatos accepted a suggestion from Zahar for change in the definition of 'novel fact': a

19. "Falsification," p. 34.

20. In *The Copernican Achievement*, ed. R. Westman (Los Angeles: University of California Press, 1975), pp. 354–83. Reprinted in *The Methodology of Scientific Research Programmes*, vol. 1, pp. 168–92. Page references are to the latter. Quotation from p. 184.

fact is novel if (although known already) it played no role in the formulation of the theory it is taken to confirm. For example, the explanation of Mercury's perihelion gave dramatic corroboration to Einstein's theory—even though as a low-level empirical proposition it had been known for almost a hundred years—because its exact solution was "an unexpected present," an unintended by-product of Einstein's program.[21]

Despite the fact that the Zahar modification seemed to be required by certain episodes in the history of science, two critics, Michael Gardner and Alan Musgrave have both pointed out that, with the new definition, assessment of theories would require often unavailable biographical information about the intentions of the scientists proposing the theories.[22] Musgrave claimed that what Lakatos really meant in his original paper was that a fact is novel if it is not a consequence of the best available background *theory*. He claimed, furthermore, that on this understanding Lakatos's definition is an acceptable one. This is too soft a criterion, however, because it fails to rule out cases where the second theory is formulated specifically to account for some anomaly with respect to the first theory *and* some already-known fact that is independent of the first. This is exactly the sort of post hoc theorizing Lakatos intended to rule out. For example, consider two versions of the theory of evolution. Specifically, let the second theory be the theory as amended by the addition of genetic theory (as an auxiliary hypothesis) to explain the variation postulated in Darwin's original formulation. Mendel's results from his experiments cross-breeding peas are a set of facts known before the amended theory was proposed and *independent of the original formulation*. But because genetic theory was developed in part to account for Mendel's results, this set of facts does not provide *novel* corroboration for the evolutionary program.

Gardner proposed that we count a theory as predicting a novel fact if it turns out to account for some fact not in its "domain" when the theory was constructed. The domain of a theory is the total body of information for which, ideally, it is expected to account. But this is not a suitable definition either, since the expectations of the scientists

21. Ibid., p. 185.

22. Michael Gardner, "Predicting Novel Facts," *British Journal for the Philosophy of Science* 33 (1982), 1–15; Alan Musgrave, "Logical versus Historical Theories of Confirmation," *British Journal for the Philosophy of Science* 25 (1974), 1–23.

at a given time as to what the theory should or should not account for are no more publicly available than is Zahar's use.

We can draw several conclusions from this discussion. First, Gardner and Musgrave were right to maintain that if Lakatos's methodology is to be an adequate historical thesis, then it is essential that the evidence for the acceptability of a theory be publicly available, and 'novel fact' must be defined in such a way as to ensure this result. Historians most often rely on documents, and therefore the definition of novelty can best be expressed in terms of what I shall call "first documentation." Second, Zahar's cases have made it clear that there must be some provision for using facts that were already known at the time a theory was proposed. But the question is how to guard against post hoc theorizing. Even if we accept Zahar's definition of 'novel' as "not used in the construction of the theory," we still need, on this account, to include some sort of time factor.

I propose, therefore, that 'novel fact' be defined as follows:

> A fact is novel if it is one not used in the construction of the theory T that it is taken to confirm. A fact not used in the construction of a theory is one whose existence, relevance to T, or interpretability in light of T is first documented after T is proposed.

This amplified definition maintains all the advantages gained by the Zahar modification yet, at the same time, it saves the methodology from the anomalous result (noted by Musgrave and Gardner) that Zahar's definition would make the rational reconstruction of the history of science depend on information not usually available to the historian. It has the further advantage of saving what was important in Lakatos's original, time-dependent, definition. His intuition that there is no instant rationality in science was correct.

So far, this new definition does nothing but save the program from a serious anomaly in a way that follows naturally from its original conception. I must, by my own account, leave it to future historians and philosophers to determine whether or not this is a progressive modification of the Lakatos–Zahar research program.[23]

23. See my article "Another Look at Novel Facts," in *Studies in History and Philosophy of Science* 20 (1989), 385–88, for details and additional references.

3.2.2 Methodology versus Historiography of Science A second area around which criticisms of Lakatos's methodology cluster is the claim that the central idea of his methodology is supported by its usefulness as the hard core of a research program in the history of science. I shall take up some of these criticisms in order of increasing seriousness.

Several readers have based their criticisms of Lakatos's methodology-as-historical-research-program on the fact that as a historical thesis it fails to measure up to some *different* view of scientific methodology. Tomas Kulka claimed that Lakatos's test of method against science fails because he has not shown that the same amount of progress (or more) could not have been attained with a different methodology.[24] This criticism is inappropriate, however, since Lakatos has in fact compared his methodology (briefly) with those of his most important predecessors, and no reasonable methodologist would demand that before a theory is accepted it be shown superior to all possible competitors. Kulka has not only missed the point that the methodology *itself* serves as a 'metamethodology' for evaluating its historical support but, furthermore, the alien criterion he has attempted to apply is something *no* theory could ever meet.

Frederick Suppe and Ernan McMullin have in effect applied a dogmatic falsificationist methodology to Lakatos's historical program. Suppe first, mistakenly, claimed that Lakatos has attempted to show that while inductivism, conventionalism, and falsificationism are all falsified by basic appraisals of the scientific elite, his own methodology of research programs "is *corroborated* in virtue of passing the same historical test."[25] He notes that McMullin has called this a most perplexing series of doctrines. It must be pointed out that Suppe has misrepresented Lakatos—and the misrepresentations indeed form a perplexing series of doctrines. Lakatos did not apply a falsificationist criterion to his own methodology. What would be the point? It only matters whether a methodology-as-historical-thesis measures up to its own criterion, since the assumption is that *it* is the correct methodology. The falsification of falsificationism is signifi-

24. "Some Problems concerning Rational Reconstruction: Comments on Elkana and Lakatos," *British Journal for the Philosophy of Science* 28 (1977), 325–44.

25. Appendix to *The Structure of Scientific Theories*, ed. Frederick Suppe (Champaign: University of Illinois Press, 1977), pp. 659–70. Quotation from p. 669.

cant, but the falsification of the methodology of research programs is not.

McMullin has also raised a more serious challenge: how "reconstructed" history can serve as evidence for a methodology.

> For if the history used in the test is reconstructed history, then the use of it to test a methodology is either question-begging or circular depending whether or not a different or the same methodology is used to effect the reconstruction. And if it is actual, unreconstructed, history then Lakatos' research-programs methodology fails to pass the test just as surely as do the other methodologies he considers; for Lakatos acknowledges that a number of the historical episodes he examines do not proceed in accordance with his research-programs methodology, although they could and should have.[26]

And Suppe added that Lakatos "appears to imply that unreconstructed history is impossible—which seems to be incompatible with [his] own practice of indicating how his own reconstructed histories deviate from the actual (unreconstructed?) history by presenting the *actual* history in footnotes. Making coherent sense of his views here is difficult, if not impossible."[27]

Thus we have two questions: the relation between reconstructed history and actual, or unreconstructed, history; and whether a circular argument is involved when one bases methodology on history reconstructed in light of that methodology. With regard to the first of these, I believe Suppe's and others' mystification could be removed quite simply by recalling Lakatos's European background. 'History' has two senses in (continental) European thought, and even two words in German. According to one usage, '*Historie*' refers to a catalog of events in chronological order. Thus *Historie* of science would list facts such as proposal of theories, results of experiments, by whom and when. The most significant facts for testing a methodology are scientists' "basic value judgments" regarding the superiority of one theory to another at a particular point in the history of science. '*Geschichte*' on the other hand refers to the narrative understanding of the human significance of events. It seeks to determine

---

26. Ernan McMullin, "The History and Philosophy of Science: A Taxonomy," in *Minnesota Studies in the Philosophy of Science*, vol. 5, ed. Roger Stuewer (Minneapolis: University of Minnesota Press, 1970), pp. 12–67. Quoted by Suppe, p. 669.

27. *The Structure of Scientific Theories*, p. 669.

human motives, causes, and so on, and knit them into an intelligible whole in such a way as to show the past's relevance for the present. Lakatos's point can be made quite nicely by substituting one or the other of these words in the appropriate place: All *Geschichte* of science involves rational reconstruction, and the reconstructions provided by various methodologies can be tested against the *Historie* of science.[28]

We can see the point of Lakatos's insistence that the historian of science needs some theory of rationality by considering a much simpler example. Suppose a professor assigns a mathematics problem, then asks a student to write the answer on the board. The class 'historian' can record the problem and the answer written by the student (and this would be *Historie*), but before *explaining* why the student wrote that answer, the historian must have an idea whether the number is the correct answer to the problem (a theory of mathematical rationality). If the answer is correct, according to the historian's theory, then the "internal history" does most of the explaining, and one need add only a few weak psychological premises (elements of external history) to complete the explanation—such as the student's desire to write the correct answer and knowledge of the relevant mathematics. If, however, the answer is wrong (in the historian's view) then an entirely different explanation will be required. Is it that the student made a computational error, did not know the principles for solving the problem, or perhaps had some motive for failing to perform as expected? Thus the theory of rationality (here mathematical; in the question at issue, scientific) is an essential precondition for knowing what can be explained on the basis of its inherent rationality (Lakatos's internal history) and what must be explained by resort to nonrational factors (the external history). We can now see the point of his stipulation that in reconstructing the history of science the program as a whole is progressive only if external history progresses whenever internal history lags. That is, whenever rational factors alone fail to account for scientific developments, their explanation must come from other kinds of factors.

28. For an alternative way to express this distinction, see Noretta Koertge, "Rational Reconstructions," in *Essays in Memory of Imre Lakatos*, ed. Cohen et al., pp. 359–69. She suggests that rational reconstructions in the history of science are comparable to Galilean idealizations of natural laws.

The second question to be drawn from McMullin's challenge is whether the argument for a methodology-as-historical-thesis is circular if the history (*Geschichte*) used to support it is reconstructed through the use of that very methodology. Notice that McMullin's objection to Lakatos's methodology at this point is based on a second-order foundationalism. He wants the data against which one tests a theory of scientific method to be independent of that theory—to be something on which all historians of science could agree, whatever their views of scientific methodology. The historicist philosophers of science have dashed all hopes for such data in science itself. It only remains for their critics to realize that such hopes are unrealizable at the metascientific level as well; the historian's data are bound to be as theory-laden as the scientist's. In fact, once we understand the role of methodology in interpreting the history of science, we can see that it would be absurd to use anything but the methodology in question. The test of a methodology *is* the attempt to use it in reconstructing history. Of course all tests are comparative, and the point is to compare methodologies on the basis of which can do a better job with the facts of scientific *Historie*.

Lakatos was not as clear as he might have been in specifying what counts as a "better" reconstruction of history. Sometimes he wrote as though it is only a matter of which methodology makes more of science look rational. But in other places he recognized that not all of science *is* rational. I suggest that the following, more guarded statement is more in line with Lakatos's best thought, though somewhat different from what he sometimes said: The best methodology is one that allows us to reconstruct as rational more of the positive value judgments of the scientific community than does any of its current competitors. That is, the more rationality that can be demonstrated the better, but only within the domain of episodes considered rational by the scientific community, with the benefit of hindsight.

As a corollary to this definition, the methodology must also provide reasons for the scientific community's judgments that certain episodes of science were clearly irrational. That is, the methodology must be demanding enough to rule out some cases. In this regard, the formulation of a methodology works like the formulation of a definition on the basis of paradigm cases. If the definition serves to describe the clear cases to which the word is applied, and to exclude clear noninstances, then it may be used to legislate on borderline cases for which our intuitions fail.

If the point of proposing a methodology that explains the history of science is to capture scientific intuitions about the best possible scientific practice, then it will not necessarily duplicate the scientists' own theories of scientific rationality. In the chess analogy, the astute observer of chess may have a better theory of the masters' intuitive strategies than any one of them could describe. The methodologies scientists claim to follow may be different from their actual practice. For example, many scientists still speak of *proving* scientific theories, even though this is impossible. Nonetheless, we should expect philosophical views of science, right or wrong, to affect practice. I suggest that the historian's knowledge of particular scientists' philosophical ideologies, when these differ from Lakatos's methodology, play the role of auxiliary hypotheses for *external* history. Uncovering them should be an important part of the positive heuristic of Lakatos's historical research program. For example, it might be shown that American theoretical psychology made very little progress (according to Lakatos's standards) in the middle of the present century, and that the (mistaken) positivist views of scientific rationality held by psychologists at that time were largely responsible for the failure.

3.2.3 Feyerabend's Critique   The most serious criticism of Lakatos's methodology is that of Paul Feyerabend, who claims that the methodology is either vacuous or will be used as an unwarranted restriction on the freedom of science.[29] His reasoning is as follows: Lakatos has said that it is rational to pursue a degenerating program *for a time* because it may later become progressive. Feyerabend points out that if there is a time limit, the methodology may force the premature abandonment of a program that would have turned around if given more time. Nonetheless, if a degenerating program may be pursued indefinitely, then there is no practical difference between Lakatos's complex methodology and Feyerabend's methodological anarchism.

I shall not attempt an answer to this criticism here.[30] It is worth noting, however, that Feyerabend takes a pragmatic approach to methodology: rules that are helpful in one situation may be harmful

29. Paul Feyerabend, "Consolations for the Specialist," in *Criticism and the Growth of Knowledge*, ed. Lakatos and Musgrave, pp. 197–230.

30. One of the best replies to Feyerabend on behalf of Lakatos, I believe, can be found in Alan Musgrave's "Method or Madness?" in *Essays in Memory of Imre Lakatos*, ed. Cohen et al., pp. 457–91.

in another. I suggest in Chapter 6 that theology's situation is so different from that of science that a restrictive use of Lakatos's methodology would clearly be beneficial for theology, even if not for most science as practiced today.

3.3 Competition for Lakatos's Methodology

The only frequently mentioned recent competitor to Lakatos's methodology is that of Larry Laudan in his *Progress and Its Problems*.[31] There is much to be said for Laudan's book: it is more readable than any of Lakatos's works and includes most of the best ideas from recent philosophy of science. The question that must be addressed here is whether his work represents any significant *improvement* over that of his teacher, Imre Lakatos.

Laudan substituted for Lakatos's notion of a research program that of a "research tradition." A research tradition consists of a number of theories, some contemporaneous and some the predecessors of the present cluster. The tradition also exhibits certain metaphysical and methodological commitments which, as an ensemble, serve to distinguish it from others. Laudan's working definition of a research tradition is that it is a set of general assumptions about the entities and processes in a domain of study, and about the appropriate methods to be used for investigating problems and constructing theories in that domain.

Research traditions are to be evaluated on the basis of their ability to generate component theories that provide solutions to an ever increasing range of empirical and conceptual problems. One appraises an individual theory by assessing the number and importance of the empirical problems it solves and deducting therefrom the number and importance of the anomalies and conceptual problems it generates. Conceptual problems can be of several types: (1) internal inconsistency or vagueness of its basic categories; (2) conflict (of varying degrees) between the theory and (a) a scientific theory from a different domain, (b) a methodological theory of the relevant scientific community, or (c) any component of the prevalent worldview. Evaluation of a theory is always comparative, never absolute.

---

31. Berkeley: University of California Press, 1977.

Laudan gives three reasons for counting his methodology superior to all previous models of theory appraisal:

(1) it is *workable*: unlike both inductivist and falsificationist models, the basic evaluation measures seem (at least in principle) to pose fewer difficulties; (2) it simultaneously offers an account of rational *acceptance* and of scientific *progress* which shows the two to be linked together in ways not explained by previous models; and (3) it comes closer to being widely applicable to the actual history of science than alternative models have been.[32]

Regrettably, all three of these claims are questionable. First, the workability of Laudan's methodology depends on the *denumerability* of conceptual problems and the possibility of giving numerical weights to their importance such that their combined weight can be subtracted from the number of solved empirical problems. It is interesting that Laudan never performed this task for even the simplest example; on the contrary, he frequently expressed his own reservations about the applicability of the procedure.[33]

Second, it is not clear what Laudan means when he claims to be the first to link acceptance of theories with progress in science, because this was one of Lakatos's most important contributions.

Third, it may indeed be true that Laudan's methodology is more widely applicable to the history of science than any other, but with all the qualifications he provides we may begin to worry whether he does not explain too much. That is, has he added so many qualifications that no episode in science, no matter how disreputable, could fail to fit?[34] Thus Laudan's own reasons for preferring his methodology seem not to hold up. So let us see if other reasons can be given.

The main differences between Laudan's and Lakatos's methodologies appear to be: (1) Laudan's definition of the research *tradition* replacing Lakatos's research program and (2) his provision for counting generation and solution of *conceptual* problems in the evaluation of research traditions. We shall see, however, that his concept of a

32. *Progress*, p. 109.
33. For example, the qualification in the passage just quoted: "at least in principle"; and p. 146: "It is thus possible, at least in principle and *perhaps* eventually in practice to be able to compare the progressiveness of different research traditions" (emphasis mine).
34. See n. 38 below and relevant text.

research tradition adds nothing of value to Lakatos's research program, and insofar as Laudan's use of conceptual problems goes beyond anything found in Lakatos's methodology, the change is undesirable.[35]

Laudan's research tradition includes ontology and methods, which together generate an evolving cluster of theories. By way of comparison, Lakatos's research program includes a hard core, which may be 'metaphysical' only in the syntactic sense of making singular existential statements, but it may also be ontological in Laudan's sense of specifying the general sorts of existents with which the auxiliary theories of the program are to concern themselves.[36] Second, the methods allowable in a Lakatosian program are specified by a combination of auxiliary hypotheses (theories of instrumentation) and the positive heuristic (general guidelines for development of the program). Third, the evolving cluster of theories in Laudan's research tradition are called by Lakatos the protective belt of auxiliary hypotheses that make empirical connections for the hard core.

Thus Laudan's research tradition differs from Lakatos's research program only verbally, not in the claims it makes about the elements ingredient to scientific research. Laudan disagreed with Lakatos, however, regarding choice of an identity- or individuating-principle for research traditions. Lakatos defined 'same research program' in terms of constancy of the hard core. Laudan chose instead to allow all parts of a research tradition to change through time: "Like Lakatos," he wrote, . . . "I want to suggest that certain elements of a research tradition are sacrosanct, and thus cannot be rejected without repudiation of the tradition itself. But unlike Lakatos, I want to insist that *the set of elements falling in this (unrejectable) class changes through time*."[37] This change makes for a gain in applicability, but at the expense of precision, and is one of several such 'softening moves' as compared to Lakatos's methodology.[38]

35. For a comparable assessment see Paul Feyerabend, "More Clothes from the Emperor's Bargain Basement," *British Journal for the Philosophy of Science* 32 (1981), 57–70.

36. For example, Cartesian metaphysics in physics and astronomy; see "Falsification," pp. 47–48.

37. *Progress*, p. 99.

38. A second example: Laudan added to his theory of the rationality of acceptance of a research tradition a theory of rationality of pursuit; thus: "it is always rational to pursue any research tradition which has a higher rate of progress than its rivals (even

The more important question is whether Laudan's use of conceptual problems in evaluation of research traditions (assuming they *could* be counted and weighted numerically) would give his methodology any advantage over Lakatos's. It would appear that Laudan's methodology is preferable in that it takes into account the question of the compatibility of a scientific theory with other parts of science and with the nonscientific parts of our current worldview. But is this the case?

First of all, Lakatos's methodology does not ignore Laudan's intrascientific conceptual problems. In discussing the grafting of Bohr's research program of light emission onto the Maxwell-Lorentz theory, with which it was inconsistent, Lakatos asserted that although consistency must remain an important regulative principle and inconsistencies must be seen as problems, it is acceptable to pursue an inconsistent program so long as it is progressive.[39] In effect, empirical progress is the primary criterion for a program, and one that is free of conceptual problems (inconsistency, either internal or with regard to another scientific theory with which it ought to be compatible) is preferable to one that is equally progressive (empirically), yet burdened by inconsistencies. Thus with regard to intrascientific conceptual problems, Laudan has not added anything not already found in Lakatos's methodology.

The final question, then, is whether Laudan has improved on Lakatos's methodology by his consideration of nonscientific conceptual problems—that is, worldview problems. Laudan claimed that it should count seriously against a theory if it conflicts with some element of the accepted worldview, including metaphysics, logic, ethics, or theology. How serious the problem is for the theory depends upon how well entrenched the nonscientific belief is and

---

if the former has lower problem-solving effectiveness)" (ibid., p. 111). This gave Laudan two different criteria to use in assessing episodes in the history of science and doubled his chances of finding that history fits his scheme. A third example: research traditions are defined as a set of assumptions about entities and processes in a domain of study and about appropriate methods to be used for investigation and construction of theories (p. 81), but there are also "nonstandard research traditions" lacking one or both of these features (p. 105). And finally, the ultimate 'softening' is to count a scientific theory as progressive under some circumstances in the absence of any empirical growth, as will be seen at the end of this section.

39. "Falsification," pp. 57–59.

upon what problem-solving capabilities would be lost by abandoning it.[40] So the crucial issue is how to unpack Laudan's two qualifiers:
the entrenchment of the nonscientific belief and its problem-solving
capabilities.

Clearly we would not want to take Laudan to be proposing that
scientific theories be required to be consistent with every well-
entrenched belief, in the absence of the latter's capacity to solve
problems, for that would tie science to past ideas whose only merit is
their age. Therefore the burden of this criterion must fall on the
problem-solving capability of the nonscientific belief. But did Laudan mean to include here its *empirical* problem-solving capability, or
merely its conceptual problem-solving ability? If he requires empirical progress, then what is at issue is really nothing different from that
of deciding between two competing scientific theories. To see this
we must return to the issue of demarcation. In Popper's work the
definitive feature of science is not its subject matter but the ability of a
theory to be confronted with and judged by the evidence. Lakatos
and most other significant philosophers of science today have followed Popper in defining science according to its method rather than
its content. Therefore, according to Lakatos, regardless of subject
matter, an empirically progressive research program *is* science. If
such a program indeed conflicts with a natural science research program, then the conflict is addressed by Lakatos's methodology as
well as by Laudan's.

Laudan maintained, however, that progress can occur *without* an
expansion of the domain of solved empirical problems, and is even
conceivable when the domain of such problems contracts.[41] It is
exactly at this point that the real difference between Lakatos and
Laudan becomes clear. In effect, Laudan's methodology counts
maintaining consistency among various nonscientific parts of the
current worldview equally important for scientific rationality as are
the discovery and explanation of new empirical facts. That is, a
scientific theory may progress, on Laudan's view, in the absence of
empirical progress by increasing its consistency with 'progressive'
nonscientific research traditions. The nonscientific research traditions in turn may be 'progressive' simply by virtue of their increasing

40. *Progress*, pp. 61, 64.
41. Ibid., pp. 68–69.

consistency with other nonscientific (nonempirical) parts of the worldview. (One might call this a methodological analogue of a coherence theory of truth.) Thus if Laudan countenances purely conceptual progress at this point, his theory provides no way to prevent the entire loss of the empirical character of science. That this is the result of his methodological theorizing can be confirmed by the fact that he has given up all attempt to offer a demarcation criterion. Although he began the book as though he intended to contribute to the philosophy of science, he says later that he was really attempting to provide a methodology that will serve as a theory of rational belief applicable to the whole of intellectual history.[42] But to explain so much may be to explain too little.

It appears that despite his own bold claims to the contrary, Laudan's work does not represent progress beyond that of Lakatos. Because no other significant competitors have appeared on the philosophy of science scene, we may conclude that for the present, at least, Lakatos's methodology stands as the best available theory of scientific rationality.

4. Consequences for Theology

Ever since philosophy of science became an independent discipline, some theologians and philosophers of religion have been concerned about its import for theology—whether positive or negative. Before we consider the applicability of Lakatos's philosophy of science to theology, let us look at a few examples of attempts to apply earlier scientific methodologies to the question of religious knowledge.

4.1 Logical Positivism

For anyone concerned about the cognitive content of theological and religious discourse, the acceptance of logical positivism was disastrous. A. J. Ayer dismissed knowledge of God in a mere two and a half pages. God cannot be known to exist on the basis of deductive reasoning because, on the positivists' view of logic, the

42. Ibid., p. 187.

proposition asserting God's existence would have to be contained in an a priori proposition. But a priori propositions are all tautologies, and no existential statement can be deduced from a tautology. Furthermore, the existence of a God such as that of Christianity cannot even be shown to be probable. The Christian claims that God is a transcendent being, not defined in terms of any empirical manifestations. Such a concept is metaphysical and, according to the verificationist theory of meaning, without any literal significance. All utterances about a transcendent God are entirely nonsensical.[43]

4.2 Falsificationism

Popper's falsificationism stirred lively debate among philosophers of religion. In the 1950s, Antony Flew rewrote John Wisdom's parable of the gardener in such a way as to raise the question whether believers do not so qualify their statements about God as to make them entirely unfalsifiable.[44] Flew, unlike Popper, took falsifiability as a criterion of meaning and intended, therefore, to call into question the meaningfulness of the claim that God exists. R. M. Hare responded to Flew's challenge, granting that the statement 'God exists' gives no information about God. Rather, such statements express the speaker's basic attitude toward the world. These attitudes, called "bliks," are unfalsifiable but nonetheless important for the way one lives.[45]

This debate raised important questions about the meaningfulness of religious language—believers must take care lest their God die the death of a thousand qualifications. It also served as a stimulus for an excellent book on religious language, Alastair McKinnon's *Falsification and Belief*.[46] McKinnon modeled his discussion of religous language on an analysis of the uses of a statement fundamental to the

43. A. J. Ayer, *Language, Truth and Logic* (London, 1936). See the Dover edition (New York: 1952), pp. 114–15.

44. The original parable is in John Wisdom, "Gods," *Proceedings of the Aristotelian Society* 45 (1944–45), 185–206, reprinted in Wisdom's *Philosophy and Psychoanalysis* (Oxford: Basil Blackwell, 1957), pp. 149–68.

45. For these and related essays, see "The University Discussion," in *New Essays in Philosophical Theology*, ed. Antony Flew and Alasdair MacIntyre (London: SCM Press, 1955), pp. 96–130.

46. The Hague: Mouton, 1970.

practice of science: 'The world has an order.' For the scientist this statement has three uses. The first is to state that the world has some specific order. Used this way, it is falsifiable, and the scientist's job is to test (attempt to falsify) particular conceptions of the world's order. The second use is self-instructive; it tells the scientist to keep looking for order. Used this way, the statement is not of a sort that could be falsified, but it is highly significant (comparable to Hare's blik). The third use McKinnon calls the ontological-linguistic use. In this sense the statement asserts that the world has some order or other. This statement is necessarily true because for the world to be a world it must possess some order, whatever that may turn out to be. The third statement is, again, unfalsifiable but highly significant.

With this analysis as a model, McKinnon turned to two examples from Christian discourse: 'I believe in God' and 'God is love.' The first uses are to make factual claims: some particular concept of God is instantiated, and God's love is of some specified kind. Used in this way, the statements are falsifiable. In fact, in the process of maturation both societies and individuals are often forced to abandon primitive notions of deity and of love. The second uses are self-instructive, meaning I must live a life of trust in God, and I must attempt to see all events as manifestations of God's love. The ontological-linguistic use commits the speaker to belief in the highest reality, whatever that may turn out to be, and to understanding 'love' as whatever characterizes that highest reality.

In the course of his work, McKinnon responded indirectly to another challenge to theism arising from a falsificationist methodology. In 1962 William Warren Bartley III had claimed in his *Retreat to Commitment* that Popper's "comprehensively critical rationalism" destroyed an important means for justifying theism.[47] Theists, he claimed, were justified in accepting a system of thought on the basis of faith so long as science and all other systems of thought also resorted to unproved foundations—the *tu quoque* argument. But Popper's methodology, rejecting the demand for indubitable foundations, defined rationality in terms of the possibility of subjecting any part of one's system to criticism. Bartley claimed that because scientists in fact do so, the *tu quoque* argument is no longer available

47. New York: Knopf.

to believers. Furthermore, if believers subjected their system to comprehensive criticism they would soon be forced to abandon it.

McKinnon has shown that the 'foundations' of both religious and scientific systems (the existence of God, the order of the world) are necessarily true within the system in question and inextricably linked with the relevant form of life (trusting God, seeking order). To borrow a term from Stephan Körner, they are "internally incorrigible."[48]

Although Popper's work has served as an important stimulus in philosophy of religion, it has received less attention from theologians. Pannenberg is one notable exception. As we saw in Chapter 2, he has made use of several aspects of Popper's philosophy, although he rejects it as a fully adequate account of scientific rationality.

4.3 Other Neopositivists

Although the works of neopositivists other than Popper have not received much attention in theology or philosophy of religion, it is obvious that much reasoning in theology takes a form comparable to what Hempel has called "hypothetico-deductive" reasoning. Theologians, as do scientists, formulate hypotheses and support them by showing that if true they would account for a given set of facts. Explanatory value is taken as confirmation, even if the form of the theory is not always a universal generalization, nor the facts strictly deducible from it, as Hempel would require. Harold K. Schilling, for example, has noted the commonalities between theology and science in this regard. In *Science and Religion: An Interpretation of Two Communities*, he claimed that it is equally true of theology as of science that it involves "the gathering of data, the formulation of concepts and general relationships, the creation and use of hypotheses, and the deductive application of generalizations and hypotheses to particular situations."[49] This is such a common form of reasoning, however, that one seldom thinks to attribute it to the philosopher who named it, whether it be C. S. Peirce's "abduction" or Hempel's more recent proposal.

48. See *Categorial Frameworks* (Oxford: Basil Blackwell, 1970), pp. 14–25.
49. New York: Scribner's, 1962, p. 37. Schilling emphasizes, however, that such descriptions of scientific method tend to hide the role of creative imagination in science.

Another theologian who might justifiably be considered here is Douglas C. Macintosh. Macintosh's *Theology as an Empirical Science*, written before the major works of the logical positivists, took J. S. Mill's methods for experimental inquiry as a guide.[50] But Macintosh advocated especially "the way up and down," which comes quite close to Hempel's hypothetico-deductive reasoning, except that philosophers since Mill have seen a much narrower role for induction. Mill's method involves three steps: "first, ascertaining the simpler laws by direct induction; second, calculating from these laws of the simpler processes what should be expected to result from their combination; and third, appeal to experience for verification of the conclusion deduced. This method might be employed in theology."[51]

More recently, John R. Carnes has argued for isomorphism between theology and science in his *Axiomatics and Dogmatics*.[52] Carnes follows Frederick Suppe in characterizing neopositivism as "the received view" and in requiring that it be updated by the addition of a realist thesis to replace the instrumentalist interpretation of theories so common among positivists. Carnes distinguishes between apologetics and dogmatics. Apologetics is based on experience and includes natural theology and the works of such theologians as Paul Tillich. In dogmatics the data are the dogmas of the church; Karl Barth serves as an example here. The point of theology of both types is to explain religious experience, which Carnes prefers to describe in Tillichian terms.

The weakness of a methodology conceived strictly in Mill's or Hempel's terms is that it offers no answer to the question how to choose between competing comprehensive theoretical structures, each of which has well-supported hypotheses. This weakness shows up in theology as well. The strength of a Hempelian starting point, of course, is that despite criticisms of his hypothetico-deductive account of theory confirmation and of his deductive-nomological account of explanation (mainly, that strict deducibility of facts from hypotheses is not possible), we still have here an important formulation of how reasoning from experience works: hypotheses are confirmed by their ability to explain observed facts.

50. See John Stuart Mill, *System of Logic* (London, 1843).
51. Douglas C. Macintosh, *Theology as an Empirical Science* (New York: Macmillan, 1919), p. 41.
52. New York: Oxford University Press, 1982.

4.4 Historicist Philosophers of Science

Thomas Kuhn's work has attracted a fair amount of attention among theologians and philosophers of religion. Basil Mitchell has compared the argument for a theistic interpretation of experience to that for a new paradigm in science.[53] Ian Barbour made use of Kuhn's work in *Myths, Models and Paradigms*, where he argued that the Christian religion is a paradigm much like those Kuhn described in science.[54] Barbour's book has been discussed and emulated by others such as Sallie McFague in *Metaphorical Theology* and Gary Gutting in *Religious Belief and Religious Skepticism*.[55] Hans Küng has used Kuhn's theory of paradigm change as a tool for reconstructing the history of theology, suggesting that the works of Augustine, Aquinas, Luther, and Calvin initiated new paradigms in Christian theology.[56]

Nicholas Wolterstorff's *Reason within the Bounds of Religion* makes use of a feature of science described by Kuhn, Lakatos, and others.[57] Arguing from the thesis that scientific theories are *underdetermined* by data, Wolterstorff urges the development of Christian scholarship wherein scientists choose from among the class of empirically adequate theories those that are most consistent with Christian convictions.

It appears that no one has tried to show the relevance of Feyerabend's philosophy for theological work. Feyerabend's thesis—that theories of scientific method are more primitive than science and would put an end not only to science but to knowledge in general when applied consistently—might suggest that the attempt to use scientific methodology in theology is misguided. He would grant, however, that because science is presently well respected in Western society and because many people believe that its essence can be captured by methodological slogans or theories, it might be an important propaganda move to show that theology, too, agrees with such slogans (theories).[58]

53. In *The Justification of Religious Belief* (London: Macmillan, 1973).
54. New York: Harper and Row, 1974.
55. Sallie McFague, *Metaphorical Theology: Models of God in Religious Language* (Philadelphia: Fortress Press, 1982); Gary Gutting, *Religious Belief and Religious Skepticism* (Notre Dame, Ind.: University of Notre Dame Press, 1982).
56. "Paradigm Change in Theology," in Hans Küng and David Tracy, eds., *Moving Toward a New Theology* (Edinburgh: T. and T. Clark, 1988).
57. Grand Rapids: Eerdmans, 1976. 2d edition, 1984.
58. In private correspondence, January 16, 1986.

4.5 Conclusions

The foregoing is certainly not a complete account of the ways philosophy of science has been found relevant to religious thought. It is adequate, however, to show a general line of development. If the logical positivists' view of science (so similar to Hume's empiricism) is taken as a standard, there is no hope for the cognitive content of theology. With the work of the neopositivists, however, the relation between theology and science again becomes a question worth investigating. With the works of the historicist philosophers of science, there are marked positive analogies between theology and science to be explored. In short, as philosophy of science has become more sophisticated and more answerable to the actual practice of science, more possibilities have emerged for showing a positive relation between philosophy of science on the one hand and philosophy of religion or theology on the other.

4.5.1 Demarcation in Question   The present chapter set out to provide an adequate theory of science—of its distinctive method of reasoning—that could then be used to test the scientific status of theology. If theology can be shown to be (either actually or potentially) scientific through the use of Lakatos's theory, this should be an interesting and important result for those theologians who, in the shadow of Hume, Kant, and others, have seen the need to be concerned about the cognitive status of their discipline.

It should also be of interest to philosophers of science whose tradition looks back to the logical positivists' attempt to formulate a criterion that would bar from the sphere of public discourse all metaphysics (and a fortiori all theology). For if the best demarcation criterion at our disposal fails to distinguish between science and theology, then the burden of proof shifts back to those philosophers who would claim that theology is not scientific, does not belong in the secular university, should not have a say in shaping public thought. In the comparison of the methodologies of Lakatos and Laudan and the rejection of the latter, we have come to see exactly what it is, in Lakatos's view, that differentiates science from other nonscientific intellectual disciplines: it is the *progressive* explanation of empirical facts. Other intellectual disciplines may produce series of theories over time, but if their aim is not to explain *empirical* facts, and if they do not, at least on rare occasions, explain such facts in a progressive manner, they are not scientific.

4.5.2 Theology as a Science Now that I have described and defended Lakatos's methodology, the time has come to raise in a preliminary way the question of the applicability of this methodology to theology. Demonstration of its applicability requires two steps: (1) to show that Lakatos's description of the structure of science is applicable to theology—that is, that there exist coherent series of theories in theology that have the formal properties of a research program; and (2) to demonstrate that occasionally some theological research programs are *empirically* progressive. If only the first of these formal conditions is met, then we can say that theology bears a more striking analogy to scientific thought than may have been supposed. But unless the second requirement is met as well, theology cannot be counted as a science.

In 1965 Claude Welch advocated the use of the term 'program' as an important descriptive device in the history of theology. One would have to hold some bizarre essentialist view of language to conclude that mere use of the word by a historian of theology shows the applicability of the same word as defined by a historian of science. Yet Welch's brief descriptions of theological programs do sound promising as candidates for Lakatosian research programs. 'Program' for him refers to "a point at which theological development has obviously come to a definite focus (often classically stated by a few individuals or even by one) [a *hard core*?]. There is a self-conscious and deliberate community in aim and method [shared *positive heuristic*?], through which the endeavors of several (and sometimes, of many) thinkers can well be interpreted."[59] Welch offers as examples of programs in theology the move toward demythologizing and existentialist interpretation, process theology, theology done with the aid of analytic philosophy, the Pannenberg group's identification of revelation as history, and the death of God theology.

In the following chapter we examine an example different from these, namely, Catholic modernism near the beginning of this cen-

59. Claude Welch, "On Theological Typology," *Theology Today* 22 (July 1965), 186.

tury, to see whether it indeed possesses the formal characteristics of a Lakatosian research program.

One should expect difficulties in finding examples from recent theology that meet the condition of empirical progress. For one reason, empirical progress is difficult to achieve even in science. Lakatos pointed out that it may take decades for a new research program to arrive at its first novel fact.[60] Alan Musgrave adds that the more progressive a program has been in the past, the more difficult it is to predict and confirm additional novel facts.[61] But there is a further difficulty to be faced in the attempt to find novel confirmation of theological programs. The predominant views stressing the 'scientific' status of theology in recent decades have been those of the history of religions school (theology is the scientific study of religions) and of the dialectical theologians (for whom theology is the 'science' of revelation). Both these approaches to 'scientific' theology focus upon types of data that are essentially historical and not repeatable.

In Pannenberg's discussion of these two options, he concluded that even though Christian theology must begin with the Christian tradition (its revelation and the phenomena of the Christian religion), it must proceed to confirmation vis-à-vis reality external to the tradition—that is, to contemporary sorts of data—if it is to fulfill its role as a science of God. The biblical concept of God as the source and end of all reality demands such a role for theology. Thus taking Scripture seriously as the starting point for Christian theology must lead to a study that seeks its data beyond the confines of Scripture.

The investigations in this chapter lead, by a very different route, to a conclusion similar to Pannenberg's: If theology is to be a science at all (in the Anglo-American sense of 'science'), then it cannot be merely a study of the Scriptures, but must seek and find some sort of grounding in contemporary empirical data (however 'empirical' may be defined). That (some) theology does (or could) meet these two requirements for scientific status cannot be known a priori; it can only be shown by examination of concrete examples. The following pages are devoted to this task.

60. "Falsification," p. 65.
61. "Logical versus Historical Theories of Confirmation," p. 21.

The Modernists—
Testing the Method

Philosophy of science without history of science is empty;
history of science without philosophy of science is blind.
—IMRE LAKATOS, "History of Science and
Its Rational Reconstructions"

I began in Chapter 1 with the thesis that theology must, if it can, substantiate its knowledge claims in the court of probable reasoning. Until it does so, Hume's agnostic conclusions stand. But as Pannenberg observes, it is now philosophy of science that provides the definitive account of the canons of probable reasoning. Consequently the appropriate response to Hume's arguments is to show that theology measures up to the best available theory of scientific method.

Recent philosophers of science have rightly grown wary of prescribing how science ought to be done without having first studied the history of science. Similar caution is in order with regard to theology. The injunction to do theology according to the prescriptions of scientific methodology is empty unless it can be shown that (some good) theology already comes close to those norms. Thus the task of this chapter is to examine a theological school (the Roman Catholic modernist movement from roughly 1890 to 1910) in order to see whether Lakatos's theory of scientific rationality allows for a reconstruction of the rationality inherent in its development. I believe we will find that, in this case at least, a theological movement bears striking formal resemblance to a Lakatosian research program. If so, a major step will have been taken toward substantiating my claim that theological knowledge can be justified in a way parallel to scientific knowledge. The remaining step will be to answer the question of what serve as data for theology, a question that will be addressed directly in Chapters 5 and 6.

Any number of modern theological movements might have been chosen to illustrate the applicability of Lakatos's scheme to theology. Nonetheless, the Catholic modernists make a particularly interesting example because they echo the thesis of Chapter 1, namely, that as scientific reasoning came to replace the medieval epistemology of authority, theology was thrown into a state of disarray. This crisis struck Catholic theology with full force nearly a century after its arrival in Protestant circles.

A second reason to begin with the modernists is that they raise for us in the clearest possible terms the question whether data can be found for theology that are not, in George Tyrrell's words, nailed down to the table by religious authority.

1. The Catholic Modernists

Roman Catholic modernism was an attempt to reconcile the Catholic heritage with the thought of the modern world.[1] The movement was provoked by the spread from Germany into broader European Catholic circles of the results of biblical and historical criticism. These discoveries made it plain to many scholars that the traditional doctrine of direct verbal inspiration of the Bible failed to account for the contents of the Scriptures, and that the church of their day differed radically from the New Testament church in both teaching and institutional structure. The latter discovery raised difficulties because of the post-Tridentine view of Catholicism as a bastion of unchanging truth.[2]

1. For history of the modernist movement see Alec Vidler, *The Modernist Movement in the Roman Church* (Cambridge: Cambridge University Press, 1934), and *A Variety of Catholic Modernists* (Cambridge: Cambridge University Press, 1970); Emile Poulat, *Histoire, dogme et critique dans la crise moderniste* (Paris: Casterman, 1962); Bernard Reardon, *Roman Catholic Modernism* (Stanford: Stanford University Press, 1970).

2. Ernesto Buonaiuti, *The Programme of Modernism*, trans. George Tyrrell (New York: Putnam's, 1908), pp. 37–38, 74–90.

One of the first to be counted among the modernists was Alfred Loisy, a French priest and biblical scholar. Loisy's life was one of conflict and controversy. He was dismissed from the Institut Catholique because of the (apparently) mistaken belief that he held views condemned in the 1893 encyclical "Providentissimus Deus."[3] He then took a post teaching in a Catholic girls' school, where the practical necessity of finding a restatement of Catholicism compatible with biblical criticism forced itself upon him. He wrote a treatise on dogmatics that was never published, although much of its content appeared in his later book *L'Evangile et l'Eglise*.[4] *L'Evangile* took the form of a Catholic reply to Adolf von Harnack's *Das Wesen des Christentums*.[5] In response to the storm of controversy it created, Loisy wrote a sequel titled *Autour d'un petit livre* in 1903.[6] In the later book's pages Harnack (1851–1930) had nearly disappeared from the scene; it was almost entirely devoted to a new interpretation of Catholicism. This was the last of Loisy's modernist writings, for by 1904 he had given up hope of transforming traditional orthodoxy and confined himself to biblical scholarship. Meanwhile the wheels of Vatican bureaucracy were turning. Loisy was excommunicated in 1908 for failing to retract his views. He lived a quiet life thereafter, until his death in 1940.

The most prolific and probably the most important theologian among the modernists was George Tyrrell. Tyrrell was born in 1861 in Ireland and raised in the Church of Ireland, but later converted to Catholicism and became a Jesuit. His writing career can be divided into three periods: an orthodox one when he wrote mostly devotional works; a "mediating liberal" phase when he was involved with the work of John Henry Newman (1801–1890); and a final, modernist period, beginning in 1900 and lasting until his death in 1909. Tyrrell left the Jesuits in 1906 because of the controversy his works had caused, and was excommunicated in 1907.[7]

3. Pope Leo XIII, in *Enchiridion Symbolorum, definitionum et declarationum de rebus fidei et morum*, ed. Henry Denzinger (Freiburg: Herder, 1965), sec. 3280ff., pp. 639–42.

4. Paris: Alphonse Picard, 1902. Translated *The Gospel and the Church* (London: Isbister, 1902).

5. Leipzig, 1900. Translated *What Is Christianity?* (London, 1901).

6. Paris: Alphonse Picard.

7. For Tyrrell's biography, see *Autobiography and Life of George Tyrrell*, 2 vols., ed. Maude Petre (London: Edward Arnold, 1912); and David G. Schultenover,

Other modernists deserve mention: in England, Baron Friedrich von Hügel was most important for the liaison work he did among others interested in the modernist cause. Maude Petre is best known as Tyrrell's friend and biographer but was not an inconsiderable theologian in her own right.[8] The Italian Ernesto Buonaiuti wrote an elegant defense of modernism and published it along with the text of the encyclical "Pascendi dominici gregis" in which modernism was condemned as "the synthesis of all heresies."[9] There were other French, English, and Italian modernists, and many sympathizers, but the movement died shortly after its condemnation and the vigorous heresy hunt that followed. I shall concentrate here on one strand of the modernist movement represented in the works of Loisy, Buonaiuti, and especially Tyrrell.

Alec Vidler, a leading historian of the movement, claimed that despite the fact that all participants shared a desire to promote the adaptation of Catholicism to the new conditions of their day, modernism was not a unified position, but rather centered around four major thrusts: (1) a response to biblical criticism and critical church history, (2) an aversion to scholasticism, (3) a revulsion from papal authoritarianism, and (4) a break with conservative political and social doctrines.[10] I hope to show here that the first three of these emphases form integral parts of a single unified *research program*. That is, if we consider both the formal requirements of a research program as defined in the previous chapter and the historical setting of Catholic modernism, understood in terms of conflict between the new scientific reasoning and the old epistemology of authority, we can see that each of these three emphases was necessary for the modernists' goal of adapting Catholicism to the new intellectual conditions. I shall not be able to examine more than a fraction of the modernists' writings here, and it would therefore be more accurate to speak of *a* modernist research program. Nonetheless, if Vidler's characterization of the major thrusts of the movement is accurate, then the

---

George Tyrrell: In Search of Catholicism (Shepherdstown, W. Va.: Patmos Press, 1981), hereafter, *Tyrrell*.

8. See Clyde Crews, *English Catholic Modernism: Maude Petre's Way of Faith* (Notre Dame, Ind.: University of Notre Dame Press, 1984).

9. The complete title of Tyrrell's English translation is *The Programme of Modernism and the Encyclical of Pius X*. The encyclical was officially published in *Acta Sanctae Sedis* 40 (1907), 593–660.

10. *A Variety*, pp. 16, 19.

program I consider here, if not *the* modernist program, must at least be one important focus of the movement as a whole.

2. The Modernist Research Program

The essential elements of a research program (from Chapter 3) include a complex series of theories where a central theory (the hard core) is unchanged while its conjoined auxiliary hypotheses are modified and amplified in order to account for an increasing domain of data. Furthermore, a mature research program is guided by a positive heuristic—a partially articulated plan for development of the protective belt of auxiliary hypotheses in the face of known anomalies and competitors. We must see whether the works of the modernists provide theological analogues for each of these elements of scientific research.

2.1 The Hard Core

I suggest that the hard core, the central theory, of the modernist program can be stated as follows:

> Genuine Catholicism is the true faith and reconcilable with modern thought.

This statement is one all the modernists considered here would have accepted as central. For example, Tyrrell wrote in his preface to the 1908 edition of Loisy's *Gospel and the Church* that Loisy would go down in history as the one who definitively stated the question "whether Roman Catholicism was compatible with the truth of Biblical and historical criticism."[11] In *Christianity at the Crossroads*, Tyrrell's last book, he wrote: "By a Modernist, I mean a churchman, of any sort, who believes in the possibility of a synthesis between the essential truth of his religion and the essential truth of modernity."[12] Buonaiuti, in *The Programme of Modernism*, wrote:

> Speaking the language of our age, and thinking its thought, we have tried to bring it into touch with the teachings of Catholicism, that

11. London: Sir Isaac Pitman, 1908, p. viii.
12. London: Longmans, Green, 1909, p. 5.

through such contact their profound mutual affinities might be made evident.

But the Church and Society can never meet on the basis of those ideas which prevailed at the Council of Trent, nor can they converse together in mediaeval language.

The pretended bases of faith [in scholastic teaching] have proved themselves rotten beyond cure. But the faith itself, that rich heritage of Catholic religious experience, we have felt beating with a new life within our hearts, and we have seen unmistakably its perfect consistence with the best exigencies of contemporary thought.[13]

The statement that I propose as the core of the program has the further value of distinguishing the modernist program from other live options of its day. The existing competitors, as the modernists saw it, were liberal Protestantism, Catholic scholasticism, and liberal Catholicism. Although liberal Protestants had worked for at least a century on reconciling Christian thought with the intellectual conditions of the modern age, their work provided no solution for the modernists, because the latter were firmly committed to salvaging Catholicism—thus the insistence that Catholicism represents the "true faith."

On the other hand there were many who had no desire to reconcile Catholicism to the modern age. Most notable among these were the Catholic scholastics who continued to maintain the adequacy of the medieval system.

Finally, there were contemporaries of the modernists, such as Wilfrid Ward, who counted themselves liberal Catholics, claimed a great debt to John Henry Newman, and worked at what the modernists believed was a mere patchwork attempt to show that the Catholicism of their day was already compatible with modern thought. I insert the word "genuine" in my statement of the core theory to distinguish the modernists from these liberal Catholics, because the modernists denied that the hierarchical institution and the predominant scholastic tradition represented genuine Catholicism, while the more conservative 'liberals' hoped to provide an apology for the church more or less as it was.[14]

13. *The Programme of Modernism*, pp. 4, 5, 7.

14. As I suggested above, there may be other modernist research programs. Gabriel Daly, for example, in *Transcendence and Immanence* (Oxford: Clarendon Press,

2.2 The Auxiliary Hypotheses

The hard core of a research program plays a quasi-definitional role in that the attempt to support it constitutes working on *that* program. It also forms the center of the theoretical structure in that all other theories of the program elaborate its meaning and relate it ('quasi-deductively') to the appropriate evidence—core theories are generally too vague or abstract to have any empirical consequences standing alone. My proposed statement of the core of this program must be shown to form such a theoretical focus. It will be identified as such if major works of the modernists serve to elaborate and justify this one hypothesis. As a matter of fact, Tyrrell, Loisy, and Buonaiuti wrote on the following topics—topics appropriate, considering their intellectual milieu, for explication of the statement I have suggested as the core of this research program.

> (1) determination of the "genuine" content of Catholicism, in contrast to Protestantism on the one hand and scholastic theology on the other;
>
> (2) determination of the meaning of 'true' in the context of "true faith";
>
> (3) determination of the relevant content of "modern thought," which included:
>   (a) contemporary empiricist epistemology,
>   (b) results of biblical criticism, and
>   (c) critical history of the church.

Several of these developments, either alone or jointly, required further elaborations of the program:

Rejection of scholastic theology and the attempt to establish a new idea of what is essential to Catholicism (1) required:

> (4) a new understanding of the nature of doctrine.

1980), suggests that the central research aim of modernism was to get clear on the relationship between the transcendent and the immanent. This may be the hard core of a separate but related modernist program. Lester Kurtz distinguishes between doctrinal modernism and philosophical modernism, so there may be two major programs involved. See *The Politics of Heresy* (Berkeley: University of California Press, 1986), pp. 8–9.

The results of biblical criticism (3b), along with rejection of the
scholastic doctrine of revelation (1), required:

> (5) a new doctrine of revelation.

The new empiricist epistemology (3a), along with the attempt to
define the meaning of "true faith" (2), required:

> (6) a theory regarding the experiential confirmation of the
> truths of the Catholic faith.

We will see how the modernists' works addressed these issues.

3. Alfred Loisy

In *The Gospel and the Church*, Loisy's concerns center around a
thesis very close to my proposal for the hard core of the modernist
program. The argument of the book can be summarized as follows:

> Central claim: Catholicism is the true heir of Gospel Chris-
> tianity (despite Harnack's claim that biblical scholarship and
> church history show otherwise).
>
> Secondary theses:
>
> (1) Catholicism today consists in rites, dogmas, and a hier-
> archical structure developed from rudimentary forms in
> the New Testament era.
>
> (2) Each of these major developments was necessary for the
> survival of the church and therefore of the Gospel.
>
> (3) But whatever is necessary for the survival of the Gospel is
> a part of true Christianity.

I suggest that these secondary theses be construed as auxiliary
hypotheses for Loisy's theological research program. Hypotheses 1–
3 together support his central theory that Catholicism is the true
faith, serving as links between his central thesis and the data collected
from among the facts of history and New Testament scholarship.
Hypothesis 3 is a methodological presupposition—that is, it determines the method by which Loisy will go about substantiating his

claim: for each of the developments he considered he would find an early form or precursor in the New Testament and then provide an argument, based upon historical evidence of changed circumstances, that the development had been necessary for the survival of the church. For example, in the case of the hierarchical structure of the church, he located its origin in the fact that Jesus called the twelve apostles, giving them authority among the other disciples, and gave Peter first place among the apostles. He then argued that because of the increased size of the church in later years and the fact that the original spirit and enthusiasm had been diluted, the church came to need more structure and discipline.[15]

So we have in Loisy's work what may be taken as one stage in the theoretical development of a theological research program that addresses some of the issues identified above as necessary for elaboration and support of the proposed hard core. Loisy began to explicate the term "genuine Catholicism" by differentiating it from Harnack's liberal Protestantism and defined some of the relevant contents of "modern thought" in his use of biblical criticism and church history. Thus *L'Evangile et l'Eglise* can be construed as an attempt to give empirical support to the theory of the reconcilability of Catholicism with modern thought by means of auxiliary hypotheses that relate the core theory to appropriate data.

4. George Tyrrell

Can one say that George Tyrrell's theological works constitute a continuation of the theological research program begun by Loisy?

A good approach to Tyrrell's development is a collection of essays that he wrote between 1899 and 1907, which thus spans nearly the whole of his modernist period. Titled *Through Scylla and Charybdis*, this volume covers a vast assortment of topics: the nature of Catholicism; the nature of revelation, dogma, and theology, and the relations among them; Newman's theory of the development of doctrine; comparison of liberal and scholastic Catholicism; mysteries; pragmatism; prophecy and history; the *consensus fidelium*; papal infal-

15. *The Gospel and the Church*, pp. 146, 149–50.

libility; and replies to critics.[16] Tyrrell claimed that the essays were placed in *logical* rather than chronological order. It will be interesting to see whether the attempt to understand his work as the development of a theological research program with the core identified above will serve to disclose the "logic" behind Tyrrell's selection and ordering of these articles. That is, do the essays contain auxiliary theological hypotheses that progressively define and support the thesis that genuine Catholicism is the true faith and reconcilable with modern thought?

4.1 Tyrrell's Hard Core: The Defense of Catholicism

The preface and introduction to *Scylla* make it plain that Tyrrell saw his task throughout to be the defense of Catholicism. These pieces were written in 1907, near the end of the period when he believed such a defense possible. Thus he expressed the challenge with heavy irony rather than the hopeful enthusiasm of his earlier modernist works. He began the preface thus:

> "I am not mad, most noble Festus," says S. Paul, "but speak forth the words of sobriety and truth." There is many a Festus to-day to whom a belief in the resurrection of Catholicism will seem the fruit either of too much learning or of too little—either of an over-crowded, over-wrought brain, or of a totally inadequate experience of men and things. (P. vii)

Tyrrell asserted that adherence to Catholicism in the face of what he granted must seem an overwhelming cumulative argument against its claims needed an explanation and, although the ordinary apologetic supports had broken down, "it may be found to have stronger and more massive pillars and buttresses than earlier apologists either needed or suspected, to which, rather than to their dialectical de-

---

16. London: Longmans, Green, 1907. Page references to this volume will be given in parentheses in the text. Other books by Tyrrell should, except for limitations of space, be considered. See for example *The Church and the Future* (London: Priory Press, 1910); *Lex Orandi* (London: Longmans, Green, 1904); *Lex Credendi* (London: Longmans, Green, 1907); and *Christianity at the Crossroads* (London: Longmans, Green, 1909; rpt., London: Allen and Unwin, 1963).

fenses, its vitality and endurance are to be ascribed" (p. 17). It was to provide just these supports that Tyrrell had drawn together this collection of essays.

4.2 Tyrrell's Auxiliary Hypotheses

4.2.1 Distinctives of "Genuine Catholicism"   I said above that the first task in explicating and supporting the hard core would be to define genuine Catholicism over against its live competitors; this is just what Tyrrell began to do in the introduction. He explained that he would *not* follow the liberal Catholics, who attempted to reconcile contemporary Catholicism both with its very different beginnings and with the contents of modern thought by means of the New-manesque theory of the development of doctrine. But it is in chapter 2, titled "Reflections on Catholicism," (written for the 1907 collection) that Tyrrell embarked on the serious task of explaining positively what he took genuine Catholicism to consist in.

Religion, Tyrrell noted, is a "natural" phenomenon, though one that results from contact with God. Christianity (the teaching and example of Jesus) is a leaven within the mass of natural religions. Jesus first criticized and purified Judaism, but his insights are equally valid criticisms of all other religions.

Catholicism includes an eclectic mixture of elements from all the world religions. Yet it is the highest religion—"the most fully developed branch of [the] tree"—because along with the vulgar, it also contains the highest attainments of the human spirit in its relation to God. It is superior to rational, purified religion ("puritanism") because its jumble of levels provides spiritual nurture to people of all sorts.

Tyrrell took great pains in "Reflections on Catholicism" to distinguish Catholicism as a historical phenomenon, "a living concrete reality," from any theory of Catholicism such as that of the scholastic apologists. The theological scheme of the church is not exhaustive of her reality; in fact, no theory could ever capture the entire phenomenon. It was confusion of the religion itself with its already dead representation in scholastic theology, Tyrrell believed, that made defense of Catholicism appear so difficult a task.

This essay continues Loisy's defense of Catholicism against Harnack's liberal Protestantism. Tyrrell retained Loisy's functional argument, that what is necessary for survival is a true development of Christianity, while amplifying the discussion of what is necessary with general considerations regarding growth of religions and cultural progress. "For in everything that has obtained wide and enduring currency in her life," Tyrrell maintained, "we recognise the product of some natural law of religious psychology, under which a discerning criticism will detect at least the perversion or imperfect expression of a vital truth" (pp. 76–77). Nonetheless, he dropped Loisy's troublesome notion that everything involved in true Catholicism in his day must have been at least germinally present in Jesus' day. Instead, he substituted the premise that Jesus' role is that of a critic or leaven for pre-existing religion. Thus Tyrrell need show not that all aspects of contemporary Catholicism were already somehow present at the beginning, but only that the purification process that Jesus brought to Judaism still continued in contemporary Catholicism. So we see here a typical move in scientific reasoning, according to Lakatos: the hard core is protected from falsification by anomalies (in this case, failure to find in Jesus' teaching primitive forms of all aspects of later Catholicism) by the substitution of one auxiliary hypothesis for another (the leaven hypothesis for the germ hypothesis).

A problem for Tyrrell, to be considered in more detail later, is his attempt (and indeed his failure) to provide a criterion for distinguishing between growth and deformation in the church. Like Loisy, he held that whatever humankind needs for spiritual growth is a legitimate development of Catholicism, but he further claimed that one can tell what has been necessary and truly "rooted in man's spiritual nature" *by seeing what survives* the sifting of the ages. Of course this makes whatever survives a valid development *by definition* and allows for no independent evaluation. In Lakatos's terms this is a content-decreasing or degenerative move because it allows for no novel or surprising outcomes.[17]

So in this chapter Tyrrell has proceeded in accordance with the

17. This is a telling criticism to raise against Tyrrell, because, as we shall see below, in his "Semper Eadem II" he criticized Newman's theory of development for a similar reason, namely the difficulty of finding a suitable criterion for distinguishing between development and deformation.

requirements for development of the hard core: he began by characterizing "genuine" Catholicism (as opposed to scholastic theology or
more narrow 'puritan' religions) and by giving reason for counting it
the true religion.

4.2.2 Empirical Method and the Truth of Catholicism   In chapter 3
we find one of the earliest of Tyrrell's modernist writings, "Lex
Orandi, Lex Credendi" (originally "The Relation of Theology to
Devotion" in *The Month*, November 1899). He said in the introduction that this essay marked the turning point in his theology and was
fundamental to all his subsequent writings.

Here Tyrrell set out to draw a distinction between theology and
revelation. The problem addressed was one the scholastics had produced by counting Scripture as "Chapter the First" of Christian
theology and then deducing further theological conclusions. By
melding Christian theology with Aristotle's metaphysics and philosophy of nature they had produced a system of thought covering
topics of all sorts, which now conflicted with much of science and
history. This created a most serious problem: If the whole system
were a logical consequence of the revealed propositions in the *deposi-
tum fidei* (to them, the contents of Scripture plus traditions handed on
by word of mouth from Jesus to the apostles and then to all succeed-
ing popes), then no part of the scholastic system could be rejected on
the basis of such conflicts. Any denial of the conclusions would work
as a *reductio ad absurdam* against the contents of revelation. Theories of
doctrinal development were being used by many in an attempt to
overcome these difficulties as they arose, but Tyrrell came to see this
as mere face-saving; he wanted a solution that went to the heart of the
problem. His distinction between revelation and theology as "genet-
ically different orders of truth and knowledge" (p. 86) was designed
to reorder the entire system of relations among theology, revelation,
dogma, and the living faith of the church, and ultimately to show
how the received theological tradition could be modified or abandoned without the true source of Christianity being abandoned as
well.

Tyrrell explained the distinction he wished to draw between theology and the "deposit of faith" by means of an analogy. Just as
scientific knowledge about nature is distinct from nature itself and
different in kind from the nature lover's direct experiential knowl-

edge of it, so theology with its abstract categories is distinct from the life of the faithful, which it is its job to explain, and different in kind from the common beliefs of the faithful, expressed in "vulgar," "anthropomorphic" language. The Scriptures—the deposit of faith—partake of this common level of expression; they are not theological treatises or propositions.

> This "deposit" of faith, this concrete, coloured, imaginative expression of Divine mysteries, as it lay in the mind of the first recipients, is both the *lex orandi* and the *lex credendi*; it is the rule and corrective, both of popular devotion and of rational theology. . . . Just as experience is the test and check of those scientific hypotheses, by which we try to classify, unite, and explain experience; so revelation is the test and check of all philosophical attempts to unify and elucidate its contents. (Pp. 95–96)

Furthermore, because devotion must also be ruled by the Scriptures it must cohere with theology. Theology "has to be reminded that, like science, its hypotheses, theories and explanations, must square with facts—the facts here being the Christian religion as lived by its consistent professors. If certain forms of prayer and devotion are undoubtedly Catholic, no theology that proves them unreal or ridiculous can be sound" (p. 104).

So in this chapter we encounter Tyrrell's main *methodological auxiliary hypotheses* and an important advance in the research program. First, he affirmed the use of empirical methods in theology. Loisy was quite explicit on this point, but it was Tyrrell who emphasized that breaking with the tradition of the past, and solving the crisis in which Catholic theology found itself, *centered* on rejecting the model of theology inherited from the Middle Ages, namely, deduction of theological conclusions from propositions in the Bible accepted on authority. Note that this position confirms Stout's thesis regarding the source of modern theology's crisis in its flight from authority.

In addition, by relating Scripture to the devotional life of the church Tyrrell provided a locus for *independent* corroboration of doctrine and theological theories. He envisioned a separate body of data, available at any historical present time, against which claims about the contents of Scripture and their consequences for theology could be tested. This move offered promise for the settling of disputes, such as that between Loisy and Harnack, where each claimed to be able to explain more of the existing data of Scripture and history.

In effect, Tyrrell *predicted* that theology could be done in a new way; theology could be construed as a set of theories to explain the experiences of the life of the church (as well as the Scriptures, the first and normative expressions of that experience). Catholic doctrines (theories) would thereby be shown to be *empirically* supported, and Catholicism, by these means, would indeed be confirmed as the true faith.

One of the functions of the positive heuristic of a scientific research program is to guide the development of a series of increasingly sophisticated models that allow for a progressively closer fit between theory and reality. Lakatos gave as an example Newton's increasingly complex models for calculating the orbits of the planets in terms of the hard core of his program—the three laws of motion and the law of gravitation. Newton began with the model of a mass-point orbiting a fixed mass, then added the complicating factors for both masses in orbit, for more planets, for the planets and sun as extended masses, then for spinning and wobbling balls, and finally for interplanetary attractions.[18] Similarly, Tyrrell's thought may be described as a series of ever more complex *models* of the Catholic Church. He considered an ever larger number of relevant factors and drew and redrew the relations among them. As I summarize each of Tyrrell's essays, then, I shall include schematic drawings to represent his increasingly sophisticated models of Catholicism.

Tyrrell was certainly aware of Loisy's model of the relation between Catholic doctrine, Scripture, and history. He did not begin *Scylla* with that model, but it will be instructive to sketch it first because Tyrrell eventually incorporated all its elements into his own. Figure 1 shows Loisy's model.

In this figure I have represented the role of data by placing them below the theoretical constructions of various sorts that are 'based upon' them; thus doctrine is based upon and explanatory of scriptural passages, which in turn are theoretical constructions (critical texts)

18. Imre Lakatos, "Falsification and the Methodology of Scientific Research Programmes," in *The Methodology of Scientific Research Programmes: Philosophical Papers,* vol. 1, ed. John Worrall and Gregory Currie (Cambridge: Cambridge University Press, 1978), p. 50.

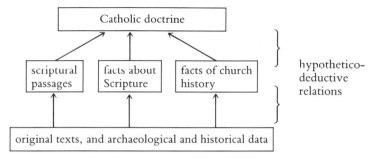

Figure 1

based on the more basic data supplied by the oldest texts and assorted information that aids in translation. "Hypothetico-deductive relations" require that the data be (roughly) deducible from the hypothesis that both explains the data and is thereby confirmed by them.

Figure 2 represents the beginning of Tyrrell's model of the church, showing the relations among theology, the devotional life, and revelation. This diagram schematically represents Tyrrell's claim that both the devotional life and theology grow out of revelation and are subject to its rule, and that there must be coherence between theology and devotion, with theology more often needing the correction of devotion than the other way round—hence the dotted line in one direction. (Arrows will be used to represent 'direction' of development of thought, but the logical relationships involved will vary, and will be noted whenever Tyrrell has specified them.)

Chapters 4 and 5, titled "Semper Eadem" and "Semper Eadem II," (from 1904 and 1905) contain criticisms of liberal Catholicism

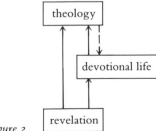

Figure 2

and scholasticism. The "development" admitted by scholastic theology (that is, explication of logical consequences) has nothing in common with the "development" of science except the name. Nor had the liberal Catholics found a way between the Scylla of the old theology and the Charybdis of scientific absolutism. Tyrrell's criticism of the liberals may be expressed as follows: they were attempting to attach an auxiliary hypothesis regarding development to the scholastic program, in order to account for the anomalies it faced, which actually *contradicted* a core theory of that program. Newman himself had soon recognized the insufficiency of the liberal doctrine of development, according to Tyrrell, and had therefore declared the need for an infallible developing authority—in effect falling into the arms of the traditionalists.[19]

In these articles Tyrrell equated the problem of finding a *via media* between liberalism and traditionalism with that of finding a *criterion* for doctrinal development and theology that neither ties theology to the thought forms of the past (the scholastic option) nor turns out to be so vague as to be inapplicable (as Newman's turned out to be).

4.2.3 Experimental Confirmation of the Catholic Faith   In chapter 6, "Mysteries a Necessity of Life" (first published in *The Month*, November–December 1902), Tyrrell turned from his attack on the competition and set about the positive task of elaborating a theory of knowledge and language that would be used to provide the criterion for doctrinal development wanted in "Semper Eadem." In this chapter he identified the *object* of religious knowledge. The world, he pointed out, is more than the natural world—the physical world is not self-explanatory. The nonphysical part of reality constitutes a "higher plane"; it is the realm of freedom, will, and love. Here he calls it the "over-natural" or "spiritual" world; in other pieces he refers to it as the "will world."

Human faculties are well suited for accumulating knowledge about nature and history, but we grasp only dimly the realities of the spiritual world. We do so by *experiment*: In building up knowledge of the spiritual world we find that

19. John Henry Newman, *Essay on the Development of Christian Doctrine* (London, 1845).

practical situations are first offered to our direct perception, as it were in their isolation, which our ignorance cannot interpret in all their connections and implications, or as to their ultimate bearings; that the Beyond to which they are felt to belong is as yet an absolute blank; that in the face of these isolated situations we determine ourselves, with prudent venture, to some kind of action or inaction, and record the observed consequences in our mind for future use; that gradually these recorded observations are systematised by the collective labour of society and pieced together into a mental construction of the spiritual world which is communicated by tradition to each member of society, to be received and modified. (P. 175)

The truth or reality of the scheme so constructed consists in its adequacy for the guidance of life, the prediction of the consequences of one's actions. The actions and experiences in question here are ethical and affective.

We have no language proper to the realities of the spiritual world and have to use the language of the natural world metaphorically. The inaptitude of language and the indirect manner of knowing combine to ensure that theories regarding the spiritual world will be less clear and adequate than those regarding the natural world. Thus such theories are called mysteries.[20] In Tyrrell's words: "It is evident that human life and progress involve an upward straining towards increasing self-adjustment with that over-natural whole of which the natural is felt to be only a part; and that this effort is futile except so far as some construction or plan of that strange country is possible; and that such a construction must necessarily be in terms of things clear and familiar, and therefore must be mysterious" (pp. 187–88).

So in this chapter Tyrrell has begun to spell out the proper domain of religious knowledge—he has distinguished between the ethical and affective experiences that form a part of the devotional life and the theories about the spiritual world that Christians devise to explain the experiences.

Figure 3 represents what Tyrrell accomplished in chapter 6. Notice that he has not yet related the work of this chapter to his original model.

20. Tyrrell's theory is much like Horace Bushnell's in his "Preliminary Dissertation on the Nature of Language, as Related to Thought and Spirit," in *God in Christ* (Hartford, 1849).

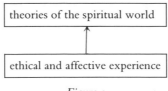

Figure 3

Tyrrell had been accused of offering a pragmatic concept of religious knowledge that was useful for apologetic purposes but evaded the tough issue of truth. In chapter 7, titled "Pragmatism" (originally "Notre attitude en face du pragmatisme," in *Annales de Philosophie Chrétienne*, October 1905), he made it clear that he was arguing for an empirical approach to religious knowledge that made increased control of experience—better adaptation to life—the test or criterion of increasing truth. Experience is not to be limited to physical sensations but includes the aesthetic, ethical, social, political, and religious dimensions of life. This clarification of Tyrrell's epistemology calls for only a minor modification in his series of models—simply to specify the logical significance of the arrow from experience to theory as that of hypothetico-deductive reasoning—as shown in Figure 4.

4.2.4 Doctrine, Revelation, Theology, and Truth In chapter 8, "The Rights and Limits of Theology" (first published in the *Quarterly Review*, October 1905) Tyrrell continued the exposition of his theory of religious knowledge, focusing on revelation and theology. What is given from "beyond" and serves as the basis for all religious knowledge is a drive or impulse to adjust oneself to the "whole" that transcends the natural world. This drive resembles an instinct and allows for recognition of courses of action or thought that move one

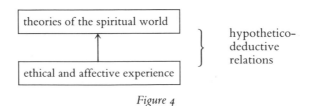

Figure 4

into closer harmony with the spiritual world. As society develops a code of behavior, sentiment, and piety on the basis of this instinct, its members reflect on this mode of life and develop "some picture, idea, and history of the world to which this code strives to adjust our conduct" (p. 207). Thus religious belief grows up *to account for* the religious life; by providing an imagined view of the spiritual world it explains religious practices and sentiments.

> So far, then, revelation (considered objectively) is a knowledge derived from, as well as concerning, the "other world," the super-natural. But its derivation is decidedly indirect. What alone is directly given from above, or from beyond, is the spiritual craving or impulse with its specific determination, with its sympathetic and antipathetic responses to the suggestions, practical or explanatory, that are pre-sented to it. . . . To find the object which shall explain this religious need and bring it to full self-consciousness is the end and purpose of the whole religious process. (P. 207)

First-order expressions of these experiences of the spiritual world and of attempts to visualize it are in poetic or "prophetic" language, which is symbolic, imaginative, imprecise. Christian dogmas are of this order. It is to be expected that forms of expression will change because in grasping for language the prophet uses whatever catego-ries, images, and concepts are available at the time.

Theological questions arise in the attempt to systematize and unify the first-order knowledge of religion and reconcile it with that from other aspects of life. As evidence accumulates, categories change and theology suffers revolutions, as do all other sciences.

Of these three elements—revelation, dogmatic expression, and theological explication—only the revelatory experience itself is guar-anteed truth. Furthermore, this truth is of a practical kind—it is truth about how to live; it is approximative and preferential—it is not guaranteed to be absolute or ultimate, but only to be the alternative that moves closer to ultimate truth.

Tyrrell explained in an introduction that he added to the article for its publication in *Scylla* that if he were to rewrite the article he would make a distinction between the apostolic revelation (expressed in Scripture) and the dogmatic decisions that grow up around it, giving special place within Christian dogma to the Scriptures: "I have since seen that the revelations of those who knew Christ personally must naturally differ in kind from later revelations, and be venerated as classical and normative" (p. 200).

In brief, the accomplishment of "The Rights and Limits of Theology" was to explicate the meaning of and criteria for 'true' as it occurs in the hard core of the program. Tyrrell also added clarifications and distinctions that aid in specifying the meaning of "genuine Catholicism," and gave a better account of the experiences that form the basis for the entire system. Theology's data are the Scriptures and dogmas of the church, which in turn are descriptions of the spiritual world devised to account for the code of Christian piety that has slowly evolved by means of a process of encouragement and discouragement by the divine spirit within.

In this chapter Tyrrell made a number of important distinctions within his model and proposed some additions. He distinguished between the experienced "yes" and "no" impulses from the Spirit, and the devotional life (including worship, sentiment, and morality) that is shaped thereby. He distinguished between dogmas (understood as imaginative pictures of the spirtual world) and theology and added the category of prophetic history (a component of the Scriptures), which is history seen from the point of view of the spiritual world. He defined theology as a theoretical enterprise designed to interpret dogma and prophetic history and to relate the knowledge contained therein to other sciences. Finally, he specified that dogmatic formulations are sometimes subject to the direct affirmation of the Spirit (as well as being invented to explain devotional practices). Given these changes and additions, Tyrrell's model can be pictured as in Figure 5.

4.2.5 Relation to Biblical Criticism and Critical Church History　A pressing problem for Tyrrell's program concerned the relation between knowledge of the spiritual world and historical knowledge, since Christianity *is* a historical religion. He addressed this issue in chapter 9, "Prophetic History" (first published in the *New York Review*, October–November 1905, under the title "The Dogmatic Reading of History"). This topic was an important element of the project of reconciling Catholicism with that very troublesome aspect of modern thought: historical and biblical criticism. He said:

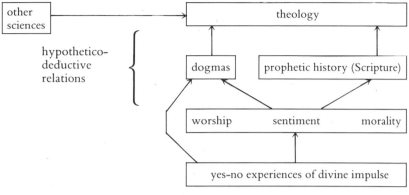

Figure 5

God has chosen to speak to us by the symbolism of deeds and facts rather than by that of visions and parables. Whence the inevitable conclusion that certain matters of fact and history, as such, belong directly to the very substance of the Christian revelation and are to be believed with divine faith, and if need be in the teeth of criticism, part of whose territory is thus brought under the distinct, and possibly conflicting, jurisdictions of faith and science. In the event of conflict, faith claims precedence; and this claim is vindicated either on "intellectualist" lines, by proving that the testimony of an omniscient and absolutely veracious witness vouches for the facts in question; or on moral lines, by showing that they are inextricably bound up with the sovereign interest of our moral and spiritual life, as part and parcel of the religion by which we live. (Pp. 243–44)

Tyrrell made three important points here. First, he reiterated that some historical facts are not just facts, but are also *symbolic* of realities of the spiritual world. Second, by mentioning and then dropping the defense of biblical history on the basis of divine authority, he separated himself from an apologetic strategy based on medieval authoritarian epistemology. Third, he suggested that some conflicts with critical history may be ignored if there is good enough reason to accept the Catholic system of which they are a part.

If Tyrrell had had the language of Lakatos's philosophy of science at his disposal, he could have put his third point in the following way: all research programs suffer from anomalies, and it is permissible, even advisable, to ignore anomalies for a time, so long as one's program is progressive over all, and one has a plan for developing auxiliary hypotheses to account for them. Tyrrell did have an auxiliary hypothesis for this purpose, which forms the substance of the article under review.

Tyrrell claimed that a mind in sympathy with the Divine can tell how history ought to go under given circumstances when it fulfills God's purposes. This sense has allowed saints and prophets in the past to fill in parts of history that are unknown through natural means, for example, the biography of Mary. Furthermore, the community interprets prophetically the 'bare facts' of history, attempting to see in them God's will and plan. Such interpretation is analogous to an artistic or dramatic rendering of an event, where the artist falsifies details in order to communicate some higher truth. The Christian tradition is history written by these means. It claims to be entirely true insofar as it gives the religious meaning of history; however, it is subject to correction in detail by the results of critical history—the actual world does not yet conform entirely to God's purposes—but only up to a point. Certain factual truths are "necessary elements of the historical core or substance of the creed" (p. 252). And it is "for the Church herself to determine what is or is not of the *substance* of her 'dogmatic' reading of history" (p. 253).

So in "Prophetic History" Tyrrell clarified the relation between the prophetic history in the Scriptures and scientific history. When the historical details are not essential to the faith, prophetic history is subject to correction by scientific history and its data. This addition is represented in Figure 6, where the broken arrow signifies the qualified use of scientific history.

In chapter 10, "The Corporate Mind" (originally "Consensus Fidelium," *New York Review*, August–September 1905), Tyrrell addressed the question of the criteria for recognizing true elements of Catholic devotion. He had said in chapter 2, "Reflections on Catholicism," that it is the staying power of a form of devotion through the centuries that provides reason for our believing it to be a legitimate part of Catholicism. But in this chapter Tyrrell distinguished his use of the *consensus fidelium* from mere universal popularity. It is not the crowd-mind (however unchanging) that represents true Catholicism, but rather the consensus of the church's active and original contributors: the saints and doctors of the church.

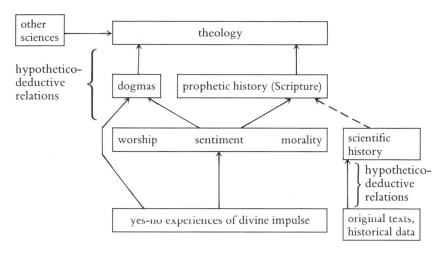

Figure 6

By this move, Tyrrell saved himself from the charge that he has made all long-lived developments in the church valid by definition. His resultant problem is that we need some *criterion* for recognizing those contributions to the Catholic mind that are higher and more saintly. Tyrrell said that the official hierarchy exists to "gather, dispense, and communise the gifts and graces of her charismatic hierarchy" (p. 262). But the question remains: how is this charismatic hierarchy to be recognized? It came to be of more than academic interest when the official hierarchy refused to recognize the modernists as saints or doctors and condemned them as heretics. Having left this task in the hands of the officials, with no criterion in his system for disputing their actions, Tyrrell would in the end find himself forced to concede that Catholicism (as it existed in his day) was not going to be reconciled with modern thought, since it refused to identify itself with the "genuine" Catholicism he envisioned.[21]

Tyrrell's work here calls for the inserting of an institutional screening process to accept some but not all of the worship, sentiment, and morality of Catholics as relevant for developing Catholic dogma and theology. Figure 7 represents this step schematically.

21. See Tyrrell's July 1908 preface to Loisy's *Gospel and the Church*, p. viii.

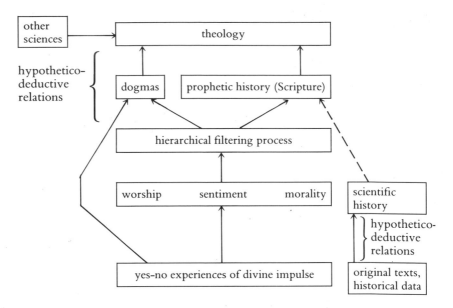

Figure 7

4.2.6 A New Doctrine of Revelation Chapter 11, "Revelation" (previously unpublished), clarified the relation between the original experience of the Divine and the written or spoken communication of that experience to others. Tyrrell specified that the word 'revelation' was to be used primarily to refer to the experience and only derivatively to denote the record or expression by which the experience is retained and communicated to others. The revelatory experience involves the whole person; it is "an experience made up of feelings and impulses and imaginings; which reverberates in every corner of the soul and leaves its impress everywhere; in the mind no less than in the heart and will" (p. 282). The mental effects are as much *given* in the experience as are the other aspects. It is possible to record these mental effects, these spontaneous expressions, and insofar as the Scriptures contain revelation this is its source. Besides records of direct mental effects of contact with God, however, the Bible contains reflections on these experiences, as well as history, theological and ethical reflections, and legends.

Tyrrell's classification of the mental element of revelation as part of the original experience allowed him to justify the move he wished to make in the introduction to "The Rights and Limits of Theology" to give special status to the apostolic revelation, which is "the vestige, the imaginative impress which Christ made on the mentality of an age that had known and seen and touched Him; that had, through Him, been brought face to face with God, and had been filled to overflowing with the Divine Spirit" (p. 291). Plainly, he said, it is of the utmost importance for the future practical and theological use of this experience that its natural self-expression be retained pure and intact as something sacred.

Apostolic revelation is final in the sense that it is normative for testing all later revelations—all later impulses attributed to the Spirit. Dogmatic decisions add nothing to the original revelation. They attempt to reassert it in other forms and to protect it by denying philosophical statements that oppose it.

Theology reflects on the expressions of the mental components of revelatory experiences, seeking to recover the whole experience from the part and to provide interpretation and understanding of the knowledge communicated therein. The church as a community has the infallible ability to recognize genuine revelation, but the theologian's reflections upon revelation—for example, discerning the implied historical and philosophical truths—are as subject to error as the work of any other scientist.

So in "Revelation" (1907?) Tyrrell amplified his account of the original revelatory experience and claimed that a mental representation or effect of the experience of contact with the Divine is inherent in the experience. He also distinguished several different components of Scripture besides records of revelatory experiences and asserted that dogma is only protective of revelation. With these changes the model may be represented as in Figure 8, where the double lines function as 'equal signs,' meaning that Scripture contains the categories so indicated; the broken line represents the testing of later putative revelations against the scriptural norm.

4.2.7 A New Understanding of the Nature of Doctrine Chapter 12, "'Theologism'—A Reply" (previously unpublished), is primarily a response to critics. A few new points deserve mention, however. First, Tyrrell furthered his argument that dogmas are merely protective of revelation. If this is the case, they should be expected to grow up as a "more or less accidental congeries of defensive propositions"

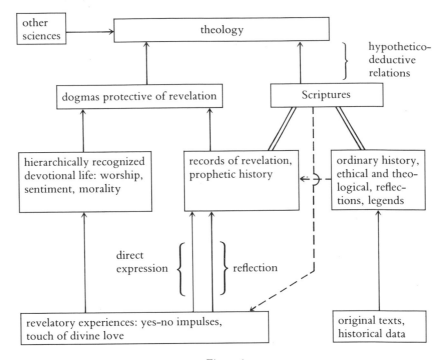

Figure 8

(p. 332) rather than in the orderly progression of deductive explication. This position, he said, accords better with the facts of history
when one examines the dogmas and the controversies out of which
they sprang. Tyrrell also expanded on the need for dogmatic pronouncements. Pointing to the change in thought forms and language
between the apostolic age and his own, he suggested that naive
reading of the Bible in light of the modern mind would result in
absurdity. The church must therefore make verbal changes to protect
the original meaning. The church does this by instinct, but critical
study of the Bible and of the historical development of doctrine also
helps protect from error by illuminating the differences between the
thought forms of these periods and our own.

   The changes occasioned in Tyrrell's model of the church were in
specifying the role of critical history of doctrinal developments in
understanding dogmatic formulations and amplifing the role of biblical criticism in understanding the Scriptures. Figure 9 shows the
resulting small change in the schema.

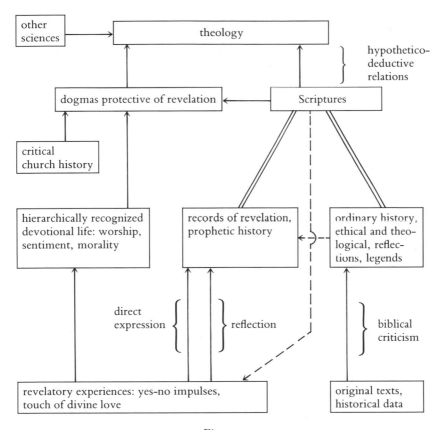

Figure 9

4.2.8 Genuine Catholicism, Concluded   In the last chapter, "From Heaven, or of Men?" (originally "Da Dio o dagli uomini?" *Il Rinnovamento*, April 1907), Tyrrell examined what he saw as an institutional error in Catholicism, the acceptance of sacerdotalism and the absolute authority of the pope, along with the theological justification of papal absolutism on the basis of a miraculous charism or spiritual gift. Tyrrell argued that both the practice and its justification erred because they conflicted with Jesus' teachings on servanthood and with the growing consensus of the faithful that God works from within his people rather than as an external miraculous agent. He also pointed out that the absolutist view of the hierarchy faced several anomalies such as clues preserved in the liturgy indicating that it is the whole community, and not only the priest, who perform the vital functions.

It is understandable that Tyrrell would invest energy in attacking this particular feature of the church, since it clashed with his theory of revelation, of the (true) rule of faith. Revelation comes to all who are disposed to receive it. The task of the hierarchy is to gather up and sift the products of this ongoing process. The First Vatican Council's theory of papal authority had arrogated to the pope alone a function that belonged to the whole church.

Tyrrell's criticism of sacerdotalism and papal absolutism suggests that the institutional life of the church (as well as its doctrine) must be consistent with the teaching of the Scriptures and must be tested according to its ability to foster the devotional life of the church. Thus Scripture, the devotional life, and church history provide data for evaluations of the church's institutional structure. The institution, in turn, provides a stimulus for theological reflection. With this addition Tyrrell's model of the church is shown in Figure 10.

4.3 The Role of Tyrrell's Models in the Modernist Program

Lakatos noted that the function of a series of models in a research program is to provide an ever closer fit between the hard core and the data upon which it is based. At this point one might well ask how the series of models Tyrrell developed in *Scylla* relate to the hard core of the modernist program: "Genuine Catholicism is the true faith, and reconcilable with modern thought." I suggest that the relation is as follows and that the whole of Tyrrell's contribution to the research program so far can be summed up very briefly in this way:

(1) The criterion for *genuine Catholicism* is that it fit Tyrrell's model.

(2) The criterion for the *truth* of Catholicism is consistency of the 'upward' development (the rest of the structure) with its data.

(3) The *reconcilability* with modern thought is built into the system as a boundary condition for theological development and dogmatic reassertion.

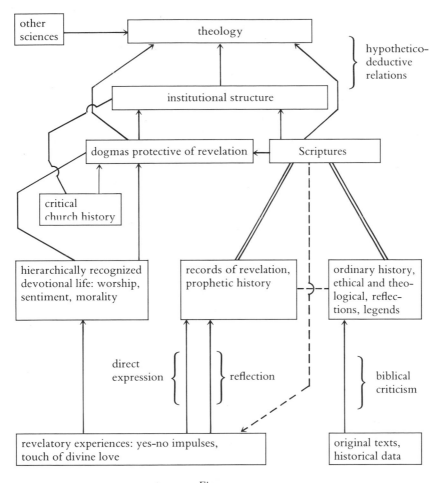

Figure 10

These three assertions are the most important of Tyrrell's auxiliary hypotheses. Notice that explication of the hard core in this manner makes it true by definition and shifts the empirical concern of the program to the question whether actual Catholicism in fact fits the model—does the reality as a whole fit the theory? Innumerable facts (about the church, its revelation, its dogmas; about ancient texts and languages; about Catholic theology and institutions, and so on) are relevant to this question. It is also obvious that each added complication in Tyrrell's model opened new areas of data for the theory; thus the program was at least theoretically progressive. The important question remaining is whether it was empirically progressive as well—that is, whether any of the novel predictions resulting from Tyrrell's theoretical development can be corroborated. This issue will be addressed in Section 6.2 below.

Of course it was not necessary to show that every detail of Catholic thought and practice accorded with the model. For one thing, all models are intended to be approximations or idealizations. For another, scientists are entitled to ignore some anomalies or treat them in an ad hoc manner so long as the program progresses over all. Tyrrell's most readily available explanation for anything that did not fit was to pronounce it a temporary aberration that would right itself in time. This was in fact his attitude toward papal absolutism during his modernist period.

5. Ernesto Buonaiuti

Buonaiuti's *Programme of Modernism* is a defense of Catholicism in light of the results of biblical criticism and early church history and at the same time a defense of modernism against the charges made in the encyclical "Pascendi dominici gregis." In general, Buonaiuti's work is consistent with the modernist program as developed by Loisy and Tyrrell. Buonaiuti, however, followed Tyrrell in rejecting what may be called Loisy's germ theory—the view that all later developments in institutional life and teaching were present in the ministry of Jesus at least in germinal form. The major part of Buonaiuti's text resembles Loisy's *Gospel and the Church* in that it reviews results of biblical criticism and critical church history, explaining how each result is compatible with his view of the church, and justifying developments by arguing they were necessary to preserve the faith in changed circumstances. Buonaiuti went beyond Loisy, however, in explaining how such results undercut the scholastic theories of revelation and of the development of the church.

The main points that Buonaiuti holds in common with Tyrrell are the following:

(1) Catholicism is treated as a human phenomenon whose development is subject to explanation in the same terms as other institutions.

(2) The Scriptures are seen to contain a mixture of religious and historical truth, the former the more important. Historical events may themselves be symbolic of religious truth.

(3) Insofar as the Gospel narratives are not history in the modern sense, they are vehicles for communicating and explaining the disciples' encounter with Christ.

(4) Dogmatic development is explanatory of religious experience and is justified insofar as it was needed in new cultural contexts.

A major difference between Tyrrell's and Buonaiuti's views lies in the account each gives of the religious experience at the foundation of the dogmatic and theological system. Tyrrell described the experience in terms of impulses from the Divine, directing the life of devotion and morality. His account was obviously intended to apply to other religions besides Christianity. Buonaiuti, on the other hand, identified the originating experience much more closely with Christ: "All these various successive and sometimes overlapping conceptions are evidently elaborated to explain that one fact of which Christian faith has continual and ever new experience, namely, that Christ lives in us and that it is He who baptises with the Holy Ghost."[22]

A clear advance for the modernist program beyond Tyrrell's work was Buonaiuti's explanation of the development of scholasticism by means of the modernists' conception of the church. Neither Loisy nor Tyrrell nor Buonaiuti was inclined to consider the possibility that the Catholic church had gone badly and permanently wrong during the course of its development. The indefectability of the church may be seen as one of the implications of the modernists' core, that genuine Catholicism *is* the true faith (and always has been and will be). Nonetheless, all three took a dim view of the scholastic system. Tyrrell reconciled the 'errors' of scholasticism with the the-

22. *The Programme of Modernism*, p. 72.

ory of indefectability by means of the ad hoc hypothesis of temporary aberrations. Buonaiuti, on the other hand, used the hypothesis, first contributed to the program by Loisy, that whatever is necessary for the church's preservation is a development of genuine Catholicism. He then described the historical situation that he believed had necessitated the development of the scholastic system in the Middle Ages. Thus scholasticism was no anomaly for the modernist program, no aberration, but rather a confirming instance for the modernist theory of church development. Nonetheless, its day had passed. Buonaiuti, at the time of his writing, maintained his optimism that the crisis would pass, and that modernist views would soon provide the next synthesis of Catholic thought.

6. Evaluation of the Modernist Program

Evaluation of a (putative) scientific theory, according to Lakatos, involves four considerations. Has it taken the form of a research program? Has it anticipated any novel facts—is it progressive? If the answer is yes to both these questions, it is scientific. (If it has the form of a research program but consistently accounts for data in a post hoc manner, it is pseudoscience.)[23] Has it developed in accordance with a positive heuristic? If so, it is mature rather than immature science. And, finally, is it more progressive than its rivals? If so, it is the theory to accept.

I believe by now it is clear that the works of Loisy, Tyrrell, and Buonaiuti constitute a developing theological tradition with all the formal characteristics of a scientific research program. The final question regarding comparison of modernism with its rivals will not be attempted here; it is not necessary for the purposes of this project to show that the modernist program was the most acceptable, but only that it is a candidate for evaluation according to the canons of scientific reasoning. Note, however, that the comparison of modernism with scholasticism would not be governed by Lakatos's methodology. Scholasticism could not be reconstructed as a research program because it did not aim at being an empirical or scientific system.

23. See Lakatos's essay on demarcation printed as the introduction to volume 1 of *The Methodology of Scientific Research Programmes*, pp. 1–7.

6.1 Had the Modernist Program a Positive Heuristic?

The purpose of Lakatos's requirement that "mature science" be guided by a positive heuristic—a plan for development of the auxiliary hypotheses—was to avoid what he called "the tacking paradox," meaning that a theory could be made progressive by having conjoined to it an auxiliary hypothesis concerning some totally unrelated subject matter. The consequences of the new auxiliary hypothesis would then become predictions of the theory as a whole, and they would be novel in the sense of not being predictable on the basis of the original theory alone. To rule out such cases of artificial progress, Lakatos stipulated that the auxiliary hypotheses should have a more intimate connection with the core theory than mere conjunction. This is the role of the positive heuristic—a coherent plan (perhaps only partially articulated) ordering the development of the program. Another way to express this: it is a plan for the development of a series of ever more complex models. Noting that some of the best scientific research programs developed this way, Lakatos made such a plan the criterion for distinguishing between mature and immature science.

We have seen that the modernists did in fact devise auxiliary hypotheses to elaborate the core theory in the ways I have recounted: defining genuine Catholicism over against its rivals, specifying the kind of truth that could be ascribed to it, and explaining how it could be reconciled with modern thought. We have also seen that Tyrrell's works can be understood as successive refinements of his model relating revelation, dogma, theology, institutions, and criticism within the church. So it remains to ask whether this work proceeded according to a preconceived plan or whether it was only a piecemeal response to anomalies and threats from competitors. I believe that the positive heuristic was sketched only several years after the program's beginning, and then perhaps only by Tyrrell.

The view that Catholicism could be reconciled with critical scholarship and the plan to do so went back at least to Loisy's tenure at the girls' school (from 1894 to 1899) and to Baron von Hügel's presentation in 1897 of a paper, "The Historical Method and the Documents of the Hexateuch," at Freiburg, Switzerland. In response to the baron's paper (heard by Loisy and read by Tyrrell), Bishop Mignot of Fréjus wrote: "You have been brave enough to say aloud what all the specialists think silently; that the conclusions of sane criticism, however bold they may seem, are not in opposition with the Faith." Tyrrell's response to von Hügel's paper, by his own account, was a sore temptation to plunge into biblical studies.[24] None of this leads one to believe, however, that a coherent strategy for theologically addressing the results of criticism existed at this time, for in the late 1890s Loisy and von Hügel were still trying to learn from liberal Protestantism, rather than to develop a rival Catholic program, and were urging the same on Tyrrell.

Only with the publication of Harnack's *What Is Christianity?* did Loisy become aware that Catholicism needed defense against the claims of liberal Protestantism. Furthermore, when he wrote *L'Evangile et l'Eglise* in 1902, he still expected the official church to see his work as a contribution to its cause. With the ensuing attack, however, he came to see the defense of "genuine" Catholicism to be defense against scholasticism as well as liberal Protestantism. Thus the sequel, *Autour d'un petit livre* (1903), called more directly for a reinterpretation of Catholicism's claims; in it Loisy addressed the important question of the nature of the truth claims made for dogmas of the church. Thus, through all of this, Loisy's modernist works were stimulated more by events without than by any preconceived plan for development of a program.

Only with hindsight did Tyrrell see "The Relation of Theology to Devotion," published in 1899, to be the first of his modernist writings. Although his early modernist works can be seen to fit into a comprehensive program, and although some of the directions it would take were predetermined by his and von Hügel's interests in mystical experience and Blondelian epistemology, it cannot be said that the program was clear in his mind from the beginning. In November 1900 he wrote to his provincial regarding a new apologetic for Catholicism: "even if I had the fullest liberty [to write and publish] I am not at all sure that I should know what to do with it; I should know the right direction but not much more."[25]

According to Tyrrell's biographer David Schultenover, Tyrrell's plan emerged only in February 1901, in response to the joint pastoral letter issued by Cardinal Vaughan and the bishops of the Province of

24. Quoted by Schultenover, *Tyrrell*, pp. 80, 81.
25. Ibid., p. 140.

Westminster (December 29, 1900). The pastoral emphasized the distinction between the *ecclesia docens* and the *ecclesia dicens*—the faithful are to accept on authority the teaching of the pope and his counselors regarding both doctrine and interpretation of Scripture. Tyrrell put the question of the interpretation of the Vatican's teaching authority in this form: Is the pope the shepherd, standing apart from the flock he leads as a being of another species, rational and active, while they are passively driven or led? Or is the pope merely the principal sheep, a member of the flock himself and spokesman for it? This was a question regarding the very constitution of the church and the very nature of the rule of faith. There was no point in discussing whether to accept more or less of the results of criticism until the place of criticism with regard to ecclesiastical teaching was settled. Thus, Schultenover says, Tyrrell outlined what would henceforth be his program—his philosophy of religion dealing with the question of the nature of faith, and his attack upon the official interpretation of the rule of faith, authority, and the source of Catholic belief.[26] Only after these questions were addressed could the proper role be determined for critical scholarship within the church. So here, at last, we can say that Tyrrell had in mind a plan for development of his contribution to the modernist program—his positive heuristic: in brief, to defend Catholicism in the face of modern scholarship by reexamining the very nature of faith, and the source and authority for Catholic belief.

6.2 Was the Modernist Program Empirically Progressive?

The modernist program showed tremendous theoretical progress. The hard core alone allowed for no empirical predictions, but as Tyrrell developed his model of Catholicism, each modification allowed for new predictions about the various aspects that constitute the institution and its teaching, and about the relations among them. For example, in the fifth version of Tyrrell's model (1905; Figure 6) he added a hypothesis about the relation of scientific history to the prophetic history in the Bible and predicted that the permanent results of scientific history will not contradict biblical history on

26. Ibid., p. 154.

fundamental points. In asserting the merely protective function of dogma (1907; see Figure 8) Tyrrell predicted that study of the history of dogmatic development would show it to have been randomly conditioned by changes in thought forms and controversies, rather than produced by orderly dialectical development. With his ninth version of the model (1907; Figure 10) and his argument that the monarchial structure of the church was incompatible with revelation and devotion, Tyrrell predicted that papal absolutism must be a passing phase—an error sooner or later to be rectified. In general, Tyrrell's theory *predicts* that dogma, theology, institutional structure, and scriptural interpretation *can all be made consistent with* the spontaneous beliefs arising from the long history of the Christian community's life of prayer and obedience.

Although a few predictions could probably be corroborated easily, it would obviously take years of work fully to evaluate Tyrrell's program, and there is a special problem in that Tyrrell, like most theologians (even those who are self-conscious empiricists), concentrated on theoretical elaboration and provided no records of the data upon which he based his theories. It goes without saying that he made no mention of novel corroboration. This latter omission is hardly surprising—only late in philosophy of science, after science was highly developed and closely competitive, was the role of novel facts ('novel' in Lakatos's technical sense) recognized in science. Furthermore, the fact that the data for theology are simply assumed makes corroboration at a later point by others nearly impossible *unless* it is from newly discovered facts. That is, we cannot provide corroboration for Tyrrell's theory that we *know* is novel using previously available facts (from Scripture, church history, or elsewhere) because we ordinarily do not know whether Tyrrell had them in mind in formulating his theories.[27] What is needed, then, is a fresh domain of data—a chance for true experimentation in an area opened up by the research program itself.

Tyrrell in fact specified a potential domain for experimental confirmation of the Catholic system in his theory of ongoing revelation. The divine Spirit's communication with individuals is private, of course, but linguistic expressions of the effects of divine impulses and

27. See my proposal in Chapter 3 for redefining 'novel fact' in terms of first documentation of the existence of a fact or of its relevance to the theory in question.

outward behavioral effects are not. If Tyrrell's theory of revelation is correct, then from the church's faithful there should be a constant supply of data regarding the *current* devotion of Catholics, which is a potential source of direct confirmation of Catholic teaching. Of course, these experiences would provide only indirect confirmation of Tyrrell's theory—it is *the fact that* doctrine can be confirmed by ongoing revelation that validates Tyrrell's model, not the content of the revelations themselves. Tyrrell's program is a theory in foundational theology or philosophy of religion, not in doctrine or theology proper.

Might there be examples of ongoing revelation (in Tyrrell's sense) in the church today? Consider the following example: Vatican II, in "The Pastoral Constitution on the Church in the Modern World," spoke of reading the signs of the times, of laboring to decipher authentic signs of God's presence and purposes in the happenings, needs, and desires in which the people of God have a part.[28] Among these signs it mentions struggles to end hunger and poverty, social and psychological slavery, political, social, economic, racial and ideological disputes, and the threat of war. Consistent with this position, the Vatican has given a great deal of attention since the council to economic and political issues in its official teaching. But is God really "the God of the oppressed"? Is there any direct confirmation today of this portrait of God as the one who struggles for human liberation? It is well known that social justice groups have grown up within the Catholic church since Vatican II. Is this only a marginal phenomenon, or is it a fact of the devotional life of the church universal and therefore a significant datum for doctrine and theology? This is certainly a question subject to empirical investigation.[29] Tyrrell's theory of Catholicism *predicts* that doctrine can be confirmed by spontaneous expressions of the devotion of the faithful. If

28. *The Documents of Vatican II*, ed. Walter M. Abbott (New York: America Press, 1966), p. 209.

29. The proper relationship between theology and the scientific study of religion, a question raised by Pannenberg, can now be addressed from the point of view of the present work. Sociologists have shown that the measurement tools of their trade are applicable to religious matters. (See for example the pages of the *Journal for the Scientific Study of Religion*.) Their techniques can be used to gather data relevant to an empirically based theology—for example, to test my suggestion that surveys would show the extent of support from Catholic devotion for the integral place of social justice concerns in Catholic piety.

empirical investigation found concern for social justice to be central to Catholic piety, this would confirm the church's teaching on God's concern for the oppressed and would provide one example of novel corroboration for Tyrrell's program.

A footnote to the Constitution's discussion of the signs of the times says: "It should be clear that not all events necessarily manifest God's true purpose. Hence the Council notes that the People of God must labor to discern the working of God's will from that of the Evil One."[30] As noted above, however, the crucial weakness in Tyrrell's system was the lack of a satisfactory criterion for recognizing *genuine* Catholic devotion. The "always and everywhere" criterion was too stringent, so he turned to the "saints and doctors" of the church but left it to the hierarchy to discern who was led and gifted by the Spirit. Does not his own tragic case show that the officials are liable to fail in this task? Without objective criteria for recognizing genuine devotion, there is no safeguard against selection of data that confirm official teaching and rejection of all that would challenge it. Discussion of criteria for judging putative revelation will be a central topic in the following chapter, along with consideration of the sense in which the devotional life of Christians can be said to provide objective knowledge *of God*.

7. Conclusions

This chapter's primary purpose was to outline a theological research program along Lakatosian lines, and I believe that this has been accomplished. The main difficulty with the modernist research program was its lack of explicit reference to the data upon which it was based, but this is a flaw in Tyrrell's style and reflects no difficulties inherent in doing theology scientifically. Appropriate data of various sorts are available—some of the most useful being recorded in the texts and practices of the Christian religion. All that remains to be done is to update theologians' theories of scientific method from those current in the early years of science to those found necessary today—to change from post hoc theorizing to progressive uses of

30. *Documents of Vatican II*, p. 209.

data. *Methodology, not subject matter,* has kept theology trailing behind in the age of science.

Finally, we return to Stout's harsh judgment on the theology of the modern era. Is it true that the epistemological shift from the age of authority to the age of probable reasoning has thrown theology and religious faith into crisis? That there was such a crisis in Catholic circles at the turn of the century is clear. I quote from a letter to Tyrrell from a fellow priest:

> May I speak quite openly to you. I was born a Catholic, and hope to die one. But I stagger like a drunkard under the theological difficulties that weigh upon my mind. It has been my lot to be thrown among intellectual men for many years now, and I need not conceal from you—I cannot conceal from myself—that to their clear-stated objections my replies have been taken as 'Chinese.' Why? Was it the fault of the defender, or the weakness of the defence? Modesty makes me fear it was the former; but if the defence was strong, why should not the defender have been strong in its strength? So I began, I was forced to begin, to ask questions of myself. . . . When such objections are pointed at me, I know not how to blunt them. . . . Difficulties are not doubts, thank God; yet they are gadflies and we would delightfully sweep them away. They are stinging endless priests now-a-days, and what these shall do when stung to madness, who shall say? . . . I fear a great outburst of rage and secessions must come on us soon, if no salve be applied. Que voulez-vous? Will men cease to think in a free land? We can no longer be convinced by being burnt. It seems to me the Church is in an evil case. She has lost the intellect of Europe; she is fast losing the intelligence. She may, like a pool, continue to hold water. But it will become stagnant water, of which men will decline to drink. . . . We ask her for bread & she gives us a stone, . . . Just a shibboleth, just a recitative, just a command, just a shrug of the shoulders, and a 'serva fidem'—nothing more. This is little help to a mind distressed. I have grown afraid even to undertake a defence of her. I enter a labyrinth with no clue in my fingers. I cannot find my way even among her own teachers. They are hopelessly entangled, ravelled like a wind-tossed skein of silk. What is wrong: If we have a deposit of revealed truth 'once communicated to the saints,' in God's name, what is it & where is it? I have asked bishops & priests & no one can tell me, and I cannot find out for myself. . . . I have found these thoughts current among priests, who, I fancied, had never thought in their lives. For the past . . . months I have been preaching up and down—, and I assure you I think I have heard the passing bell of a

dying creed. This is strong language; but the passing bell rings very loud in my ears everywhere.[31]

Tyrrell attributed this crisis to the fact that the theology being taught in seminaries was from another age:

> When a fairly intelligent priest happens to come in contact with modern doubt, the fabric of his seminary theology collapses like a card-house. Why so? Because though it means much in "Chinese," it means nothing in English. He has no interpreter's key; no bridge from one world to the other; & is scared, confused & panic-stricken. And this is because one uniform seminary course is imposed on the whole Church without respect to varieties of period, language, nationality; because the world has been developing one sort of culture; & we another; so that mutual understanding has become impossible.[32]

The "Chinese" system is theology based on authority, expanded by means of deductive reasoning. The English bishops' joint pastoral of 1900 emphasized exactly these points. It opened with a broadside against liberal Catholics who "take leave to discuss theology and the government of the Church with the same freedom of speech and opinion" as they use in other matters, but with the evil intent to reject the authority of the Church and to substitute the principle of private judgment for the principle of obedience to religious authority.[33] Of course "religious authority" is not the same as epistemological authority, unless religious authority is equated with divine authority in matters of faith. But this is exactly the equation reasserted by the joint pastoral: "The doctrines of faith . . . have been entrusted, as a Divine deposit, to the teaching Church, and to her alone—to guard faithfully, and to develop and explain, with Divine and infallible authority. Truths, therefore, at one time held implicitly, by degrees become explicitly realised and defined."[34]

Tyrrell recognized that authority had lost its sway over the minds of his fellow Catholics. Of the officials he said: " 'Authority' is their one note—their whole tune. They do not see that it is a question-

---

31. Quoted in Schultenover, *Tyrrell*, pp. 141–42.
32. Letter to Tyrrell's prior, quoted in ibid., p. 142.
33. Ibid., p. 145.
34. Quoted in Tyrrell, *Autobiography*, vol. 2, p. 151.

begging note; that it is the existence, or at least the right limits of authority, which has been brought into question."[35]

Because the age of faith had passed, Tyrrell saw his day to be much like that of early Christianity "when the Gospel had to be made compelling in a context of unbelief and even hostility. Then it was not the 'persuasive words of human wisdom' but a certain show of power that won the crowd—experimental, not rational, proof of the Gospel."[36] That Tyrrell saw the task of theology in the new era to be a scientific explanation of the experiential life associated with the Gospel has, I hope, been made clear enough in the preceding pages.

Thus Tyrrell's assessment of the source of the crisis in the Catholicism of his day foreshadowed Stout's. His response and that of the other modernists, however, was not by any means to resort to fideism or to cognitive evacuation of the words of the creed, but to meet the challenge of the new epistemology head on. Just as Pannenberg has done, Tyrrell and company set out to reform theology along scientific lines in order to make it intelligible to people in the age of probable reasoning. The modernists did not succeed as well as we might have hoped, for they, like Pannenberg, labored under the handicap of an inadequate theory of scientific method. But they have missed the mark by so small a margin that Stout's pessimism about theology's future seems premature, to say the least.

35. Quoted in Schultenover, *Tyrrell*, p. 148.
36. Schultenover's summary of Tyrrell's "Tracts for the Millions," *The Month*, November 1900, *Tyrrell*, p. 135.

[5]

Data for Theology

> All religious experience is material for the psychology of reli-
> gion; [which] has no criterion for distinguishing between true
> and false religions; it cannot say the first thing about the exis-
> tence or nature of God. . . . The scientific theologian, there-
> fore, will have to select from the manifold of religious
> experience those elements which give knowledge of God,
> just as the physicist selects from the multitude of the ele-
> ments of sense-experience those which are of importance for
> the understanding of the nature of matter and energy. The
> theologian must therefore not only have access to religious
> experience; he must have the proper means for distinguishing
> the divine.
>
> —D. C. MACINTOSH, *Theology as an Empirical Science*

The most pressing requirement in a search for suitable data for theology, as Macintosh points out, is to find ways to distinguish data that have a bearing upon the nature of God from those that bear only on the psychology (or history) of religion.

By the end of this volume we shall have noted several different kinds of facts that may provide suitable data for theology: scriptural texts, historical facts, sociological or anthropological data, perhaps even facts from the natural sciences. It is not my role here to legislate on the kinds of facts that may be used to support theological research programs. The categories of appropriate data must be determined by the content of the research program itself, in particular by its positive heuristic and by auxiliary hypotheses of a methodological nature. Nonetheless, this chapter will pursue one possible (and, I believe, very promising) source of data for theology.

We begin our quest with Tyrrell's suggestion that facts of the devotional and moral life of the church can, under certain circumstances, function as data for theology. Insofar as theological doctrines (theories) are needed to account for these facts, the latter provide data that serve to confirm the theories. Tyrrell focused on the problem of how to recognize genuine Catholic devotion (and we would certainly want to broaden his question to that of genuine or authentic Christian devotion). But another question is more pressing. How can the devotional practices of Christians tell the theologian anything about God? In Tyrrell's view the two questions are intimately related: genuine Catholic (Christian) devotion is a *response* to the actions of God in the lives of Christians—it is directed by impulses from the Holy Spirit. Insofar as devotion and morality reflect the intentions and actions of God, they provide evidence to support theories about the nature of God.

So we begin by considering two interrelated kinds of phenomena. On the one hand, there are various kinds of encounters with the triune God. In the development of biblical religion, encounters with God or with the risen Christ have often been seen as face-to-face, and in that sense external encounters, while dealing with God's Spirit is often spoken of as an inward dealing. Theophanies being rather rare, our attention here will focus on the work of the Spirit within. On the other hand, there are observable events in church life and enduring practices and convictions that arise as a result of encounters with God.

Consider this example. A common devotional practice of medieval Catholicism was the offering to God of one's patiently endured sufferings for the sake of one's own salvation or that of others.[1] Furthermore, the practice was thought to be commanded by God, both in Scripture and through assorted visions and locutions. For example, the fourteenth-century mystic Henry Suso reported a vision wherein an angel warned him of impending suffering. Henry cried out to God either to spare him or to explain the reason for the suffering. A voice within urged him to take comfort, for "I myself will be with you to help you overcome these calamities by my grace." Later, when Henry complained that his trials were too arduous, he was reminded that his reward would be an eternal one.[2]

Now, such a practice became highly controversial in postreforma-

1. Richard Kieckhefer, *Unquiet Souls: Fourteenth-Century Saints and Their Religious Milieu* (Chicago: University of Chicago Press, 1984), p. 89.
2. Ibid., p. 58.

tion Christianity, because its endorsement has obvious implications for the doctrine of salvation (and indirect implications as well for much of the rest of a systematic theology). What would it take, therefore, to show this to be a genuine datum for theology, and not just for a phenomenology of medieval Catholicism? It is clear that Tyrrell's stated criteria are inadequate to answer this question satisfactorily. In the fourteenth century the practice met both of his (successive) criteria for genuine devotion: it was widely enough accepted to appear to satisfy Vincent of Lerins's "always and everywhere"; and it was practiced and encouraged by hierarchically recognized "saints and doctors" of the church. But neither of these facts is likely to impress a reformed theologian.

Tyrrell gives us a more helpful hint, however, in his suggestion that genuine devotion is that which occurs as a result of impulses from the Holy Spirit. Therefore the key question in deciding what to make of the practice itself is whether we can make anything of the associated claim that it is a result of the action of God. *Was* the voice that Henry heard the voice of the Spirit of Christ? How is one in general to distinguish between religious experiences that represent encounters with God or impulses of the Spirit and those that do not? How is one to recognize the voice or action of God? If these interrelated questions can be answered, then the theologian will know what to make of the practices so inspired—whether or not to count them as evidence for the character and intentions of God.

Fortunately the Christian church has provided a rich treasury of answers to just these questions. Such answers fall under the headings of "discernment," or "Christian judgment," or "testing the spirits." In the following sections we survey various periods of Christian history and works originating in three major traditions (Jonathan Edwards, the Protestant; Ignatius Loyola, the Catholic; and Pilgram Marpeck, the Anabaptist). From this survey we will be able to draw together a general set of guidelines for communal Christian discernment. With these guidelines in hand one can go on to talk about events in the devotional lives of Christian communities as inspired by the Holy Spirit and therefore as acts or words of God.

I shall propose that the crucial data for theology are the results of Christian discernment. Such results will include judgments about devotional and moral practices, namely that they are (or are not) inspired by the Spirit. But the judgments may relate God's activity to other kinds of phenomena as well. Thus my category of data is more inclusive than Tyrrell's. It is also more restrictive, however, in that it excludes from consideration devotional and moral practices that cannot be shown to be Spirit-inspired. But, before pursuing the *results* of Christian discernment, we must first consider the practice itself.

1. Jonathan Edwards

We turn first to the American theologian of the Great Awakening, Jonathan Edwards (1703–1758), of whom it has been said that "the whole of his thought might be viewed as one magnificent answer to the question, *What is true religion?*"[3] Despite differences in time, setting, and religious background, Tyrrell might have made good use of Edwards's works, for they had a great deal in common. Both favored a scientific approach to theology and an experiential form of religion. Edwards had a special interest in "religious affections" and, like Tyrrell, he insisted on close relations among feeling, thought, and will. For Edwards, affections are "the more vigorous and sensible exercises of the inclination and will of the soul."[4]

Tyrrell and Edwards both claimed that by virtue of the action of the Holy Spirit true Christians have the ability to recognize what is or is not of God. For Tyrrell the impulses from the Holy Spirit to do or abstain from certain actions or forms of devotion are the basis from which a code of piety develops, which in turn serves as a source of data for doctrinal theology. Similarly, Edwards argued that when the Holy Spirit takes up residence in Christians' souls they receive a new sense—a new faculty for judging and willing—that allows them to perceive the spiritual beauty of God and thence to judge all conduct according to whether or not it partakes of God's beautiful holiness. In effect, the true Christian acquires an attraction to that which is of God and a revulsion from things that are in opposition to God (*Affections*, pp. 206, 281–82).

Nonetheless, both Edwards and Tyrrell, with their pastoral expe-

3. John E. Smith, in Introduction to Edwards's *Treatise Concerning Religious Affections*, in *The Works of Jonathan Edwards* (hereafter *Works*), vol. 2 (New Haven: Yale University Press, 1959), p. 2.
    4. *A Treatise Concerning Religious Affections*, p. 96. Hereafter, *Affections*.

rience and knowledge of the history of Christianity, recognized the possibility of counterfeit religious affections, of illusions and delusions, and therefore of the need for criteria for testing the spirits. Their concerns here progressed in parallel fashion, both beginning with the problem of recognizing impulses (or affections) truly of the Spirit of God and later focusing on the recognition of true saints. Tyrrell could have profited immensely from Edwards's careful observations and detailed study of the Scriptures regarding these matters.

1.1 Edwards's Observations

In the preceding chapter I faulted Tyrrell for failing to record the data upon which his theological theories were based. No such complaint can be made against Edwards. Two of his works are particularly important in this regard, recording observations of the effects of revivals. *A Faithful Narrative* described the "frontier revival" in his home church in Northampton, Massachusetts, and in the surrounding county beginning in 1734. *Some Thoughts concerning the Revival* was written in 1742 to convince the skeptical that the Great Awakening, throughout all of New England, was a work of the Spirit of God.[5] Here he included a great deal of descriptive material reporting the effects he had observed: numerous conversions, growing interest in religion, increased fidelity to the Christian moral code—in general, a "glorious alteration in the town." He also provided several biographical sketches. Most interesting was an account of changes in the life of an anonymous congregant whom we know to have been his wife, Sarah Pierpont Edwards. He described Sarah's increasingly intense religious experiences, her improved emotional health, and her newly felt "universal benevolence to mankind, with a longing as it were to embrace the whole world in the arms of pity and love."[6]

After his general description of the effects of the revival, Edwards

5. Both appear in *Works*, vol. 4, ed. C. C. Goen (New Haven: Yale University Press, 1972).
6. See *The Works of President Edwards*, vol. 1, ed. Timothy Dwight (New York: Carrill, 1830) for a more detailed account of Sarah's experience; and James McClendon's *Ethics* (Nashville: Abingdon, 1986), chap. 4, for the importance of this experience for Jonathan's mature theology.

gave an account of the regularities (laws) and variations he had observed in converts' experiences. "I therefore proceed," he said, "to give an account of the manner of persons being wrought upon: and here there is a vast variety, perhaps as manifold as the subjects of the operation; but in many things there is a great analogy in all" (*Narrative*, p. 160). Those "wrought upon" are first awakened with a sense of their sinful condition and their "danger of perishing." This awakening may happen suddenly or gradually. The immediate effects involve avoidance of sinful practices (many long-standing feuds ended during the revival) and adoption of the "means of grace" (reading, prayer, and the ordinances). The period of fear usually begins with a sense of outward sins, later moves to a sense of sins of the heart, and ends with a conviction of the sinner's utter helplessness and willingness to accept God's justice.

After the period of fear comes one of comfort: the truths of religion seem new; converts' hearts are drawn in love to God; their minds are filled with contemplation of the glory of God. Grace produces modesty and meekness, a concern for salvation of sinners, and great affection for others, along with confession of and reparation for past injuries to them. For some it is very noticeable when the changes first began, though for others it is not. Edwards noted that these changes tended to endure even after the revival ended.

It has been suggested that religious experience, as data for theology, is less objective than scientific data because it is more subject to suggestion.[7] Perhaps sensible of this danger, Edwards noted that the people of Northampton had earlier formed notions of how conversion was supposed to take place, yet during the revival their actual experiences were such that they found themselves "much confounded in their notions and overthrown in many of their former conceits" (*Narrative*, p. 174).

Some Thoughts concerning the Revival contains a list similar to that in *Narrative* of the positive effects of the Great Awakening: a great increase in serious consideration of things of the eternal world; awakening of conscience, awareness of sin, and reform of behavior; a new delight in reading the Bible and church attendance, along with a new conviction of the truth of Christian teaching; a new love for and

7. For example, Ian Barbour in *Issues in Science and Religion* (New York: Harper and Row, 1966), p. 258.

delight in God, and love of others resulting in confession of faults and restitution. All these phenomena, Edwards wrote, had lasted more than a year and a half (*Some Thoughts*, pp. 325–30).

At the same time, along with the positive effects he noted, a great deal of foolishness and emotional frenzy accompanied the revivals in some places. So the question how to distinguish between the work of God and other, merely human, phenomena became a pressing one.

1.2 Criteria for Judging Religious Experience

Edwards set forth criteria for judging religious life in two works: *The Distinguishing Marks of a Work of the Spirit of God*, based on a lecture given in New Haven in 1741,[8] and his culminating work, *A Treatise concerning Religious Affections*, first published in 1746. Although it was written later, we begin with *Affections*.

1.2.1 Criteria for Recognition of the Saints   Edwards intended his "twelve distinguishing signs of truly gracious affections" to offer means for distinguishing "saving" works of the Holy Spirit from "common" works of the Spirit (i.e., the Holy Spirit operating on but not dwelling in the soul) and from other influences—in short, means to distinguish whether a person is a true saint or not. Edwards drew his marks from Scripture, but clearly it was Scripture read through eyes that had long observed the effects of religious excitement.

It is important to note that the genuine signs of the Spirit's saving activity are not related extrinsically or accidentally to sainthood; thus we do not merely *infer* the presence of the Spirit from presence of the signs. Yet manifestation of these marks is not identical with sainthood; there is more to sainthood than can be observed. To see the intrinsic relation here between sign and thing signified is to understand a significant part of Edwards's theology. For Edwards, salvation or sainthood was equivalent to the indwelling of the Holy Spirit, which produces a new nature and creates a new sense of the beauty of God. The Holy Spirit, acting consistently within the saints, causes them to assume the character of God in human form—in other words, to be changed into the moral and affective likeness of Jesus. These characteristics are known as the fruits of the Holy Spirit. For

8. *Works*, vol. 4.

the most part the twelve positive signs aim to spell out the consequences of this new creation.

In his introduction to *Affections*, John E. Smith says that Edwards left vague the exact relation between sign and sainthood. I suggest that the proper technical language for the relation is to call these signs *defining criteria* of sainthood. Ludwig Wittgenstein used this expression to refer to observable signs of inward processes that are neither identical with those processes nor mere symptoms of them, but are rather so closely related that a true sentence of the form 'Sign *s* is a criterion of process *p*' is tautologous—for example, 'cries of pain are a criterion for being in pain' is tautologous. Nonetheless, the necessary relation between the meaning of the term referring to the process (e.g., 'pain' or 'sainthood') and its criterion (e.g., cries of pain, or perseverance in the Christian life) does not imply that the existence of the process can *necessarily* be inferred from the presence of the sign in particular cases; pain and sainthood can both be mimicked. But without the validity of various criteria we would not know *how* to mimic them.[9]

Although Edwards presented the signs as though they comprised a list of independent entries, I believe they can better be read as a progressive unfolding of his doctrine that true religion (true sainthood) consists in acquiring and manifesting the character of God as a result of the indwelling of the Holy Spirit. Thus some of the signs are not themselves observable but instead constitute steps in the unfolding of this doctrine *on the way to* making explicit the observable effects of sainthood. Accordingly the first sign deals with the source of truly gracious affections, namely God.

> I. Affections that are truly spiritual and gracious, do arise from those influences and operations on the heart, which are *spiritual, supernatural* and *divine*. (P. 197)

The second through fourth signs expand on this first point—on the source of gracious affections and the means by which they are caused.

> II. The first objective ground of gracious affections, is the transcendently excellent and amiable nature of divine things, as they are in

9. For a summary of Wittgenstein's views here and of the controversy regarding their interpretation, see Anthony Kenny, "Criterion," in *The Encyclopedia of Philosophy*, vol. 2, pp. 258–61.

themselves; and not any conceived relation they bear to self, or self-interest. (P. 240)

III. Those affections that are truly holy, are primarily founded on the loveliness of the moral excellency of divine things. Or (to express it otherwise), a love of divine things for the beauty and sweetness of their moral excellency, is the first beginning and spring of all holy affections. (Pp. 253–54)

IV. Gracious affections do arise from the mind's being enlightened, rightly and spiritually to understand or apprehend divine things. (P. 266)

To paraphrase Edwards so far, truly gracious affections begin with disinterested contemplation of the intrinsic beauty and moral excellence of God and of divine things, occasioned by enlightenment of the understanding.

It is only with the fifth sign that Edwards began to consider observable effects by which to judge the source of affections.

V. Truly gracious affections are attended with a reasonable and spiritual conviction of the judgment, of the reality and certainty of divine things. (P. 291)

VI. Gracious affections are attended with evangelical humiliation. (P. 311)

VII. Another thing, wherein gracious affections are distinguished from others, is, that they are attended with a change of nature. (P. 340)

VIII. Truly gracious affections differ from those affections that are false and delusive, in that they tend to, and are attended with the lamblike, dovelike spirit and temper of Jesus Christ; or in other words, they naturally beget and promote such a spirit of love, meekness, quietness, forgiveness and mercy, as appeared in Christ. (Pp. 344–45)

IX. Gracious affections soften the heart, and are attended and followed with a Christian tenderness of spirit. (P. 357)

X. Another thing wherein those affections that are truly gracious and holy, differ from those that are false, is beautiful symmetry and proportion. (P. 365)

XI. Another great and very distinguishing difference between gracious affections and others is, that gracious affections, the higher they are raised, the more is a spiritual appetite and longing of soul after spiritual attainments, increased. On the contrary, false affections rest satisfied in themselves. (P. 376)

XII. Gracious and holy affections have their exercise and fruit in Christian practice. I mean, they have that influence and power upon him who is the subject of 'em, that they cause that a practice which is universally conformed to, and directed by Christian rules, should be the practice and business of his life. (P. 383)

The key point here is the change of nature that is essential to sainthood (seventh and eighth signs). Edwards spoke of it as the acquisition of a new faculty of understanding and will, and in these other six signs he spelled out the consequences of the new, Christlike nature: conviction of the truth of one's new understanding of divine things, humility before God and neighbor, a softened heart, affections that are neither self-satisfied nor out of balance, and finally, that issue in consistent Christian practice.

Notice that with the exception of the twelfth sign even these come closer to definitions of holiness than visible signs of it, for we feel tempted in each case to ask what are the *signs* of a real change in nature, of true humility, of a softened heart. It is understandable, then, that Edwards described the twelfth sign as the sign of signs, for it is in the end the measure of all the rest, and the only one that is readily visible to another.

1.2.2 Distinguishing Marks of a Work of the Spirit of God "There is no question whatever," said Edwards, "that is of greater importance to mankind, and that it more concerns every individual person to be well resolved in, than this, what are the distinguishing qualifications of those that are in favor with God, and entitled to his eternal rewards?" (*Affections*, p. 84). Indeed, this may be the most significant question one may ask for one's own sake. For the sake of the public and communal task of theology, however, it is less important than the question of how we are to know about God himself. For this purpose the more important question may be: "How do we know what God is doing?" or another way of expressing it: "How do we distinguish between the action of God in human life and other influences?" Answering this question—giving guidelines for trying the spirits—had been the task of Edwards's earlier book, *The Distinguishing Marks of a Work of the Spirit of God*.

The question regarding the action of God is not unrelated to that of the status of one's salvation, for God's saving acts in individual hearts are of central importance. Edwards was careful to point out, however, that even the most eminent saints still manifest some degree of corruption, they still sin, and therefore their actions, thoughts, and affections cannot automatically be identified with the action of God. Thus for the present purpose of finding suitable data for theology, we need public criteria, not for judging our own status in God's eyes but for discerning what is or is not an effect of the operation of the Spirit. To know God we must be able to distinguish God's acts from those of his sinful creatures.

Edwards began *The Distinguishing Marks* with a quotation from the First Epistle of John:

> Beloved, believe not every spirit, but try the spirits whether they are of God; because many false prophets are gone out into the world. (1 John 4:1)

He emphasized here, as he would later in *Affections*, that Scripture alone should be the guide in such matters. Thus Edwards drew his five signs from an analysis of the remainder of this chapter of the Epistle. As in *Affections*, he began with a list of "negative signs"—not signs that the Spirit is not at work but, rather, characteristics some in his day believed to be signs of the Spirit but which, in fact, were not reliable indicators one way or another—for example, effects on people's bodies or imaginations. The five positive signs are as follows:

> 1. When that spirit that is at work amongst a people is observed to operate after such a manner, as to raise their esteem of that Jesus that was born of the Virgin, and was crucified without the gates of Jerusalem; and seems more to confirm and establish their minds in the truth of what the Gospel declares to us of his being the Son of God, and the Saviour of men; 'tis a sure sign that that spirit is the Spirit of God. (P. 249)

> 2. When the spirit that is at work operates against the interest of Satan's kingdom, which lies in encouraging and establishing sin, and cherishing men's worldly lusts; this is a sure sign that 'tis a true, and not a false spirit. (Pp. 250–51)

> 3. That spirit that operates in such a manner, as to cause in men a greater regard to the Holy Scriptures, and establishes them more in their truth and divinity, is certainly the Spirit of God. (P. 253)

> 4. If by observing the manner of the operation of a spirit that is at work among a people, we see that it operates as a spirit of truth, leading

persons to truth, convincing them of those things that are true, we
may safely determine that 'tis a right and true spirit. (P. 354)

5. If the spirit that is at work among a people operates as a spirit of love
to God and man, 'tis a sure sign that 'tis the Spirit of God. (P. 255)

This last sign Edwards took to be most important for, with the
author of the Epistle, he asserted that love is the very nature of the
Spirit of God. Love makes the soul long for God and Christ, quells
contentions, gives peace and good will, excites to outward acts of
kindness, and causes delight in those that appear to be children of
God. And, finally, the surest sign by which to distinguish truly
divine love and its effects is humility. Here, as in *Affections*, the last
and most important mark must itself be judged of by means of its
proper perceptible signs.

So in Edwards's works we find the claim that Christians are able to
apply a set of criteria to an assortment of religious phenomena in
order to recognize those that are caused by the activity of the Holy
Spirit. To support Edwards's claim, and to broaden our understanding of the criteria and how they are applied in practice, we turn to
others who make similar claims.

2. Ignatius of Loyola

Ignatius (1491–1556) began writing his *Spiritual Exercises* in 1523,
two years after his conversion, but made significant additions and
changes throughout his life as spiritual director and founder of the
Jesuit order. In their final form, the *Exercises* present a program for a
month-long retreat offered to followers who desire direction for
determining their vocation or state of life.[10] But the pattern of interaction with God expected to develop during the retreat bears a
striking resemblance to the conversions and rededications Jonathan
Edwards observed during the revivals—a combination of activities
or "exercises" such as Edwards would advocate as "means of grace,"

10. For a good English translation and contemporary commentary, see David L.
Fleming, *The Spiritual Exercises: A Literal Translation and a Contemporary Reading* (St.
Louis: The Institute of Jesuit Sources, 1978). Citations to the Fleming volume will
hereafter appear in the text, followed, where applicable, by a reference to the para-
graph number in the *Exercises*.

along with effects worked upon the retreatant or convert by God. The first week emphasizes meditation on one's sinfulness and on the horror of sin; the second week, contemplation of the justice, glory, and goodness of God. The third week focuses on the life of Jesus and God's plan for salvation through Christ, while the fourth week concentrates on recognition of God as the source of justice, goodness, pity, and mercy. The retreatant seeks from God the gifts of sorrow for sin, resolve to live a life in God's service, freedom from excesses in personal habits, the ability to relinquish desires for wealth and honor, and most important, humble love and gratitude to God. In short, the pattern here, as in Edwards's church, is conviction, consolation, and a new life lived in imitation of Christ.

2.1 Rules for Discernment of Spirits

The interesting part of the *Exercises* for present purposes is Ignatius's section on discernment of spirits, entitled "Rules for Perceiving and Knowing in Some Manner the Different Movements Which are Caused in the Soul." Ignatius's translator David Fleming explains that 'spirits,' as used in these guidelines, might be understood as "movements of one's heart or spirit," "motions affecting one's interior life," "a certain impetus in one's life," or "a feeling for or against some course of action" (p. 202). Note the similarity between these "movements" or "spirits" and Edwards's "affections"—changes of heart that involve judgment and impetus to act. Just as the entire notion of the thirty-day retreat expresses an activist attitude toward the Christian life, the rules for discernment of spirits, as they have come to be called, are also oriented toward practice. Ignatius offered them in order that the retreatant might accept and cooperate with impulses from God and reject the evil ones.

Ignatius supplied two lists of rules for discernment: one for the first week of the retreat, when the retreatant is struggling with sin, and another for the second, intended for making more subtle judgments. In contrast to Edwards, who concentrated primarily on recognizing the actions of the Holy Spirit, Ignatius gave more attention to the nature and operation of evil spirits.

In the first set of rules Ignatius explained that the Spirit and the enemy use opposite strategies depending on whether one is moving from bad to worse or in the direction of greater holiness. The enemy encourages the sinner to continue by offering pleasures to the imagination, but for the saint "it is the way of the evil spirit to bite, sadden and put obstacles, disquieting with false reasons, that one may not go on" (p. 204; para. 315). On the other hand, the Holy Spirit deals gently and sweetly with the saint but pricks the consciences of sinners through the process of reason. Notice that Ignatius's rules for recognizing the works of the Holy Spirit (or evil spirits) presuppose what Edwards concluded in *Affections*, namely that saints are "visible" (Edwards's term); they can ordinarily be distinguished from sinners by the observable character of their lives.

Two important terms for Ignatius are 'consolation' and 'desolation.' 'Consolation' refers to an

> interior movement in the soul . . . through which the soul comes to be inflamed with love of its Creator and Lord; and when it can in consequence love no created thing on the face of the earth in itself, but in the Creator of them all.
>
> Likewise, when it sheds tears that move to love of its Lord, whether out of sorrow for one's sins, or for the Passion of Christ our Lord, or because of other things directly connected with His service and praise.
>
> Finally, I call consolation every increase of hope, faith and charity, and all interior joy which calls and attracts to heavenly things and to the salvation of one's soul, quieting it and giving it peace in its Creator and Lord. (P. 206; para. 316)

Notice that the experiences Ignatius referred to as consolations match almost exactly those Edwards labeled "truly gracious affections." 'Desolation,' on the other hand, refers to "darkness" and "disturbance" of soul, "movement to things low and earthly, the unquiet of different agitations and temptations, moving to want of confidence, without hope, without love, when one finds oneself all lazy, tepid, sad, and as if separated from his Creator and Lord" (p. 206; para. 317). Edwards alluded to similar conditions among those not yet converted, and also used the expression 'low in grace' of those who suffered discouragement and anxiety after conversion. It is apparently the latter condition that Ignatius had in mind, and while Edwards mentioned it only in passing, Ignatius took it to be a very important fact about the course of Christians' spiritual lives.

Rules One through Four of the first week deal with the foregoing material. Rules Five through Nine give advice to those suffering desolation—for example, that one should not change a course of action determined upon during consolation and thus under the guidance of the Holy Spirit. Rules Ten and Eleven give advice to those in consolation—especially enjoining humility. The three final rules describe the usual tactics of evil spirits in their attempts to attack and delude souls.

The first two rules for the second week emphasize that it is a work of God to bring consolations to the soul. One should expect the operation of the Holy Spirit when love of God arises without any cause besides God. The remaining six rules aid in recognizing subtle clues to the sources of impulses. The enemy can counterfeit the experiences of consolation for his own purposes—namely to lead into sin—and can insinuate evil intentions into a course of thought that had its inception in God's action. One must resist this trick by recognizing the evil to which it leads. Also, the purely natural thoughts and inclinations that occur after a time of consolation must not be attributed to the action of God.

2.2 Edwards and Ignatius—Substantial Agreement

As I mentioned briefly above, Edwards and Ignatius describe similar patterns in Christian experience: conviction of sin; sense of God's holiness; the turning of one's life to imitation of Christ and service of God; overwhelming sense of God's love; and new love of God and neighbor.

Both see the Holy Spirit as the source of "consolations," or joyful "affections," and also as the source of impulses guiding behavior. Both emotions and impulses, however, can be counterfeited, so there is further need to judge their sources.

For Edwards, the works of the Holy Spirit are to be recognized by the following signs: that they raise esteem for Jesus, work against Satan's kingdom and sin, produce a greater regard for the Scriptures, confirm the truth of Christianity, and inspire a spirit of love, which is to be recognized by its effects. It produces a longing for God and Christ, quells contentions, brings peace and good will, inspires kindness, produces delight in the children of God, and is characterized overall by humility.

Summarizing Ignatius's rules, we might say that when consolation comes (meaning: love of God and of creatures in God; tears of sorrow for sin and esteem for Christ's passion; hope, faith, charity, and joy; attraction to heavenly things; and quiet and peace) in the context of meditation on the Scriptures, especially the life of Jesus and the holiness and love of God, then the putative consolation may be assumed to be a work of the Spirit of God so long as it is accompanied by humility and not shown to be counterfeit by its role in leading to sin.

Although the arrangement is different, Ignatius's work clearly parallels and thus supports, across an otherwise wide theological chasm, what we have already seen in Edwards's, namely that Christians (pastors, theologians) are able, under certain circumstances, to distinguish from among human actions those that are also the acts of God. For a complementary approach to Christian judgment let us now turn to the Anabaptist, or Radical Reformation, tradition.

3. Judgment in the Anabaptist Tradition

We have considered important resources concerning Christian judgment or discernment from two of the major Christian traditions in the West. The sixteenth-century Anabaptists and their heirs form a third major tradition.[11] The first Anabaptists were followers of the Protestant Reformer Ulrich Zwingli. These disciples began to contest Zwingli's leadership in 1523, when he decided to refer plans for church reform to the Zurich city council. The radicals originally wanted only to carry out more quickly and thoroughly the reforms already begun, but the points of greatest contention ultimately came to be the separation of church and state and the restriction of baptism to confessed believers—hence the pejorative name 'anabaptists' or 'rebaptizers.' Despite severe persecution the movement survived. Its most numerous direct heirs are the Mennonites, but similar views on scriptural authority, evangelism, baptism, discipleship, congrega-

11. For history of the Anabaptist tradition, see Donald F. Durnbaugh, *The Believers' Church: The History and Character of Radical Protestantism* (New York: Macmillan, 1968); Hans J. Hillerbrand, *The Reformation: A Narrative History Related by Contemporary Observers and Participants* (New York: Harper and Row, 1964), pp. 214-97; and George H. Williams, *The Radical Reformation* (Philadelphia: Westminster Press, 1962).

tional church polity, and separation of church and state are now represented in a large number of Christian churches.

For the Anabaptists the principle of consistency with Scripture (particularly the New Testament) took such a prominent place in their decision-making that discussion of these matters by contemporary historians often falls under the topic of "anabaptist hermeneutics."[12] Judgment always took place in the context of a need for guidance in questions concerning a local congregation's practice. John Howard Yoder points out that since the Holy Spirit was promised to the church in the context of the reconciling approach to wayward members (Matt. 18:19–20; John 20:22–23) the mandate and enablement to discern the will of God were seen as provisions not primarily for scholars, but for the concrete congregation struggling with differing visions of what obedience meant in its own time and place. This is not to say that scholarship had no place in the community; only that it must be relevant to shaping the community's common life.

The Anabaptists called on the congregation to make decisions in three types of cases: public disputations, synods, and discipline of individuals. We see an example of the first in Balthasar Hubmaier's debate (1524) with the Catholic John Eck. The key provision in Hubmaier's proposed rules for the disputation was drawn from 1 Corinthians 14:29 ("Let two or three prophets speak, and let the others weigh what is said"). The debate would take place before a congregation gathered for worship. The congregation would listen as the two men expressed their convictions, and then decide who had spoken more nearly in accord with the Scriptures.[13]

Perhaps the most significant synod in the earliest years of the Anabaptist movement was the meeting at Schleitheim, on the Swiss-German border. This meeting took place in February 1527, only two

---

12. See, for example, J. H. Yoder, "The Hermeneutics of the Anabaptists," *Mennonite Quarterly Review* 41 (October 1967), 291–307; William Klassen, "Anabaptist Hermeneutics: The Letter and the Spirit," *MQR* 40 (April 1966), 83–96; and William Klassen, *Covenant and Community: The Life and Writings of Pilgram Marpeck* (Grand Rapids, Mich.: Eerdmans, 1968).

13. The rules proposed in Hubmaier's *Axiomata* (1524) are cited in Yoder, "The Hermeneutics of the Anabaptists." For a translation of all of Hubmaier's writings see *Balthasar Hubmaier*, trans. H. Wayne Pipkin and John H. Yoder (Scottdale, Pa.: Herald Press, 1989).

years after Anabaptism became a self-conscious movement. Already there were grave threats from without—repression by the Protestants had just reached the level of capital punishment—and from within, for the young leadership met with difficulties in maintaining agreement within the ranks. The meeting at Schleitheim produced a document containing the Seven Articles, which can best be read as a list of the features distinguishing the Anabaptists from the magisterial reformers: (1) believer's baptism, (2) in accordance with Matthew 18, exercise of the ban (i.e., exclusion of offenders from the church) as the ultimate act of discipline, (3) breaking of bread reserved for the congregation united by believer's baptism, (4) separation from the world, (5) shepherds (pastors) chosen from among the congregation, (6) rejection of the sword even if authorized by law, and (7) refusal of all oaths.[14] These seven articles were agreed upon after prayer and discussion, despite the fact that "some brothers previously had understood [them] wrongly and in a way not conformed to the true meaning." The fact that the representatives of the various communities had been "brought to agreement in the Lord"[15] despite previous disagreements was taken as a sign of the work of the Holy Spirit among them. The articles were all amply referenced to the New Testament by means of marginal notes.

The Anabaptists practiced the ban, even to excess in some communities, but that excess was criticized within the movement.[16] Apparently the practice of discussion between the community and the wayward individual was often effective in creating unity in thought and lifestyle.

In summary, we can see that whereas consistency with the Scriptures served as the *criterion* for decision-making among the radicals, the *practical test* of consistency was the agreement of the entire community—whether the issue was the conduct of an individual in the local community, the distinguishing features of the Anabaptist movement as a whole, or a theological debate with outsiders. The *means* of reaching agreement was open discussion within the context of prayer.

14. J. H. Yoder, *The Legacy of Michael Sattler* (Scottdale, Pa.: Herald Press, 1973), pp. 28–54.

15. From Michael Sattler's cover letter, ibid., p. 42.

16. See Klassen, *Covenant and Community*, p. 97.

These are the general features of Anabaptist views on judgment. I turn now to one Anabaptist writer who dealt with the question of how to recognize the leading of the Holy Spirit in teaching the Bible.

3.1 Pilgram Marpeck's Criteria for Judging Teachers

Pilgram Marpeck (1495?–1556), an important leader of the Anabaptist movement in the south of Germany, wrote extensively and participated in several public disputations. Like the others, he was committed to the *sola Scriptura* principle, but without the aid of the Holy Spirit, he maintained, the Scriptures could not be interpreted or explained. The most subtle temptation besetting the Christian is to ascribe to the Holy Spirit what is actually one's own human opinion. "Ah, my brethren, how diligently and carefully we have to take heed that we do not consider our own impulse the impulse of the Holy Spirit, our own course the course and walk of Christ."[17] Therefore he offered four signs by which to judge the impulses behind one's own or a fellow Christian's teaching:

> First is love for God and to grant to my neighbor that which God has granted and given me for His praise and the salvation of my soul. Second is a devaluation and giving up of life unto death to suffer for the sake of Christ and the Gospel and all patience. Third, to realize when God unlocks or opens a door that one may enter the same with the teaching of the Gospel. No one shall open a door which God has not opened, in order that the office of the Holy Spirit remain His own and free. For He it is who opens and no one closes, He closes and no one opens, in order that the pearls not be cast before swine. . . . Fourth, that one be free and sound in teaching and judgments and in truth, in order that none speak unless Christ work through his Holy Spirit. . . . These four parts are the true proof that the compulsion is of the Holy Spirit; also that it brings forth fruit at each season.[18]

Just as Edwards formulated his signs of gracious affections by relating the scriptural norms to events he observed in his own time, so Marpeck's signs reflect his milieu. The emphasis on love was a

17. "Two Letters by Pilgram Marpeck," *Mennonite Quarterly Review* 32 (July 1958), 198.
18. Ibid., p. 199.

reflection of the teachings of Luther and of the *Devotio Moderna*, as well as of the Bible. The emphasis on willingness to suffer was aimed at the spiritualists who avoided persecution by teaching that outward forms were insignificant. The third sign, the openness of the audience, reflected his own missionary career. The fourth criterion, the freedom of the teacher, arose from Marpeck's reaction to teaching distorted by the attempt to maintain church-state unity.[19]

3.2 A Complement to Edwards and Ignatius

Among the Anabaptists we see even more dedication to thinking with the word of God in the Scriptures than we found in Edwards and Ignatius. We see again the emphasis on love as a sign of God's work and among the Anabaptists an even greater emphasis on reconciliation—"we have been brought to unity" (*Vereinigung*) rang through their documents. Note Marpeck's addition of "fruit" as a criterion for judging the Spirit's inspiration of teaching.[20]

We see again a connection between discernment of the Spirit and recognition of sin in individual life, although the connection works differently in each case. For Edwards, only the saints can recognize a work of the Holy Spirit. Ignatius had two sets of rules for discernment—one for the first week when retreatants still struggled to be free of sin, and another for later. For the Anabaptists, mutual recognition of sin and the community's call to make things right constituted the paradigm for all other discernment and action.

With the Anabaptists the practical turn we saw in comparing Ignatius to Edwards is greatly magnified, and Edwards's preoccupation with ascertaining the state of one's own soul is entirely lacking. Willingness to suffer for Christ takes a much more prominent place.

Notice, also, that for the Anabaptists the Spirit's role in shaping the practices and doctrine of the community is much more creative than it was for Edwards or Ignatius. We see the development of practices new in their day such as believer's baptism, the ban, and rejection of the sword and oaths. And whereas for Edwards the

19. Klassen, *Covenant and Community*, p. 76.
20. Perhaps '*auch*', translated as 'also' in the last line of the quotation above, would be better rendered as 'namely', indicating that "fruit" sums up the signs.

Spirit's teaching function is limited to *confirming* those convictions that his tradition takes to be consequences of revelation, the Anabaptists see the role of the Spirit being to help them *determine* (afresh) what the Bible teaches.

Both Ignatius and Edwards presupposed that the Holy Spirit works in Christian community. Edwards set forth criteria for recognizing a work of the Spirit among a people, although he applied it to individual cases; Ignatius presupposed a Catholic ecclesiology wherein the Holy Spirit works through the church. It is especially among the Anabaptists and their heirs, however, that we find the belief in the Holy Spirit's primary relation to a community expressed in communitarian practices. The Anabaptist practice of communal discussion corrects for individual delusion and ignorance.

Therefore I conclude that this tradition offers a significant complement to the works of Ignatius and Edwards on discernment of spirits, both because of its communal dimension and because of the more creative role of discernment in shaping Anabaptist practices and convictions.

4. Judgment in the Early Church

It is not possible within the scope of the present work to give an exhaustive account of discernment practices throughout the history of the church. My survey so far involves three major Christian traditions, but touches only three centuries. For a slightly more balanced account, let us consider judgment practices in the earliest years of the church.

Bernard Cooke points out that in the New Testament judgment belongs to God and the risen Lord alone, yet "we are again facing that enigmatic situation which is the essence of mankind's religious behavior. . . . From one point of view God does everything, man is totally dependent; yet, human activity is not just illusion, it does somehow contribute to a joint divine-human effort."[21] Thus legitimate human judgment must express the judgment of God. Cooke

21. *Ministry to Word and Sacraments* (Philadelphia: Fortress Press, 1976), p. 405. For a complementary account of New Testament belief and practice, see Wayne A. Meeks, *The First Urban Christians: The Social World of the Apostle Paul* (New Haven: Yale University Press, 1983).

sees judgment in the New Testament period pertaining to three matters: judgment on the truth of faith and revelation, judgment on the value and correctness of human behavior, and judgment on sin— or in other words, doctrine, ethics, and discipline.

Judgment on the truth of doctrine originates with the alteration of human understanding that comes with God's revelation in Jesus. Christians must judge the accuracy of the community's understanding of this revelation. In the early years the first-hand witnesses had special authority, and in later times the apostolic literature was given a normative role in the judging of the community's understanding. Yet the entire community participated in judging the truth of teaching because both prophets and teachers needed the approval of the community. Nonetheless, the community's judgment was in fact to be that of the Holy Spirit, bearing witness to the truth of the Christian gospel. When the Spirit came to converts they received "no infusion of ideas . . . instead there would be an internal power of awareness which would 'sense' the incompatibility of certain notions with the gospel of Christ."[22] This illumination of mind and heart effected direct insight into truth as well as direction into a way of life that led indirectly to a greater ability to perceive the truth. Thus, besides agreement with the apostolic witness, a second criterion was to be found in the deeds of the teacher—"Thus you will know them by their fruits" (Matt. 7:20).

The community also made judgments regarding behavior—both before the fact in its ethical teachings and after the fact in its judgment on sin. Here, as in judgment on truth, the process had to be communal and undertaken with the guidance of the Holy Spirit. Community leaders, in setting out guidelines for conduct, sought the direction of the Spirit—"For it has seemed good to the Holy Spirit and to us" (Acts 15:28). The procedures of Matthew 18 for use in the case of offenders involved the entire community. Prophets played a special role in passing judgment on the activity of the community and on the meaning of its religious understanding by virtue of the special charism or gift of discernment of spirits that they possessed. But again, the community bore final responsibility for recognizing genuine prophets.

In the immediate ante-Nicene period Christians continued to ex-

22. Cooke, *Ministry to Word and Sacraments*, p. 407.

pect that the Holy Spirit (or the Logos) would enable them to discern truth from falsity in faith and practical judgments. The church maintained the criteria of conformity with the apostolic witness and of fostering peace, love, and unity, but the question of who could make such judgments came increasingly to the fore. And, increasingly, the answer given was the *episkopos*. By the high Patristic period, however, interest in the bishop's ability to judge, on the basis of personal sanctity and special gifts of the Spirit, began to be overshadowed by the question of his right or power to judge, based on his office. Cooke, a Catholic theologian, suggests that this has been true, too true, of much of Christian practice since.[23]

In summary, judgment in the earliest years of the church combined emphases found in the later traditions examined above. Taken together, these lead to the conclusion that an adequate list of criteria for discernment in the Christian church includes the following: (1) agreement with the apostolic witness, (2) production of a Christlike character in those affected, meaning specifically freedom from sin and manifestation of the fruits of the Spirit, and (3) unity within the community based on prayerful discussion.

5. Contemporary Forms of Christian Judgment

So far we have found agreement that Christians are able, because of the indwelling of the Holy Spirit, to recognize what is or is not a genuine work of the Holy Spirit, whether it be a matter of teaching or of practice. We have also found a great deal of agreement on criteria for making such judgments. I turn now to some contemporary communities, not only for the sake of historical breadth, but because in some cases the contemporary authors give a more detailed account of the practices of communal judgment.

In the Catholic church the thirty-day retreat is still in use for lay as well as ordained and religious members. Other forms of retreat employ adaptations of Ignatius's Spiritual Exercises. Many religious communities and some Catholic parishes practice communal discernment in making decisions. There are various formats, but in general the process involves prayer for the leading of the Spirit,

23. Ibid., chaps. 21–22.

discussion, and reflections on Scripture, leading to formation of consensus.

For those heirs of the Radical Reformation tied most directly to the original Anabaptists, practices of communal discipline and decision-making have remained largely unchanged. In 1959 Anabaptist historian Franklin Littell, addressing the Thirty-fifth Session of the General Conference Mennonite Church, traced "the apostolic pattern of decision-making"—study of the Bible, prayer, and discussion issuing in consensus—from the Anabaptists in Zurich to today's free churches.[24]

The Society of Friends is considered by many to be heir to the Radical tradition, although only indirectly tied to the Zurich reformers. Yet from George Fox to the present, discernment and communal decision have been marks of a distinctive way of life. According to Howard Brinton, a Quaker worship meeting is meant to be a time of waiting on the Lord to speak.

> As the worshipers sit together in silence to wait upon the Lord, anyone among them may find arising in his consciousness a message which he feels is intended for more than himself alone. . . . He must learn to recognize the unique sense of urgency which is evidence of a divine requirement. If a thought comes to him with peculiar life and power, he may be justified in assuming that this is a sign from God to speak. He may sometimes be mistaken. There is no sure test of divine guidance in this or any other undertaking. If, however, through prayer and humble waiting he has become sensitive to the "still, small voice," he will be increasingly enabled to recognize a call when it comes. He will learn to recognize and reject the wish to speak when it comes from a different source.[25]

Other clues to the source of an impulse are a sense of "burden and uneasiness" before speaking and a feeling of complete peace following obedience to a genuine call. Note that if we are taking the deliverances of communal discernment as potential data for theology, we find another, more primitive level of data underlying the judgments reached by the community—the individual's "sense of urgency," or the manner in which a thought arises.

24. Franklin Littell, "The Work of the Holy Spirit in Group Decisions," *Mennonite Quarterly Review* 34 (April 1960), 75–96.
25. *Friends for 300 Years* (New York: Harper, 1952), p. 84.

A strong inward sense that some action should be taken may arise in the mind of any individual Quaker. Decisions to implement such actions are made in business meetings. These meetings must be preceded by a period of worship in which the "hard shell of ego-centricity" is dissolved and the group becomes a community. After prayer, the item of business is presented and "talked up," with no one excluded from the conversation. If serious difficulties arise, the group returns to prayer. Ordinarily a decision is made only when the community reaches complete agreement. (It took a full eighty years for American Quakers to come to agreement concerning slavery!) It is surprising, Brinton claims, how often real unity is reached even when sharp divisions were apparent, and often the accepted solution follows lines not even considered at the beginning. The Quaker believes that a deeper truth is always possible, so even one's most cherished convictions should be held humbly, and each speaker should be credited with at least some insight into the truth. Further-more, the meeting should hesitate to accept any suggestion that runs counter to the accumulated wisdom of the saints and prophets who have gone before.

As did the Anabaptists, Quakers take the emergence of consensus as the sign of a work of the Holy Spirit in their midst. Brinton recognizes a problem here, in that there is such a thing as a "lower unity"—the agreement among a group of bandits. He answers that prayer and earnest effort to purge out selfishness will lead to a higher level. Jonathan Edwards posed a similar problem in his ethical writings—what to make of the "secondary virtue" exemplified in the mutual loyalty of a band of pirates. For Edwards true virtue—a higher unity—can be distinguished from secondary virtue on the basis of the unrestrictedness of its benevolence.[26]

We find a similarly detailed description of communal discernment in the contemporary Charismatic Renewal, found today in the Catholic church and in all major Protestant denominations.[27] It became a recognizable movement in 1960, appearing in American Catholicism in 1966. The movement owes a great deal to 'classical' pentecostal-

26. See Edwards, *The Nature of True Virtue*, ed. William K. Frankena (Ann Arbor: University of Michigan Press, 1960), originally published in 1765.

27. See the April 1975 issue of *New Covenant* magazine for a series of articles on community discernment.

ism, which appeared in the United States at the beginning of this century.[28] The Charismatic Renewal differs from the older pentecostal movement in several ways, and I attend only to the later movement.

Kevin and Dorothy Ranaghan describe experiences in the renewal that closely resemble those Edwards reported during the revivals in his own day:

> In general, we all experienced and witnessed in each other the breakthrough of the love of Christ in our lives. . . . Many were drawn to long periods of prayer, marked by the predominance of the praise of God. Some found themselves opening the bible anew with a real hunger for the word of God. Just about everyone found a new boldness in faith, a desire to witness about Jesus. . . . Divisions, even hatreds, between brothers were healed.[29]

The name 'Charismatic' comes from the Greek word *charisma*, used in the New Testament, especially by Paul, to refer to gifts of the Holy Spirit. The most common examples are found in 1 Corinthians 12:8–10:

> To one is given through the Spirit the utterance of wisdom, and to another the utterance of knowledge according to the same Spirit, to another faith by the same Spirit, to another gifts of healing by the one Spirit, to another the working of miracles, to another prophecy, to another the ability to distinguish between spirits, to another various kinds of tongues, to another the interpretation of tongues.[30]

Charismatics understand a (genuine) gift to be the Holy Spirit operating in and through a person's words or deeds, for the sake of the community. The Spirit guides and empowers but does not overpower, so a gift requires the intelligent cooperation of the one who exercises it. Consequently the action or message involved in a gift is at once the action or word of God and of the person who exercises the gift.[31]

28. Arthur C. Piepkorn, *Profiles in Belief: The Religious Bodies of the United States and Canada*, vol. 3 (New York: Harper and Row, 1979), p. 89.

29. *Catholic Pentecostals* (Paramus, N.J.: Paulist Press, 1969), pp. 40–41.

30. See also 1 Cor. 12:28–30; Rom. 12:6–8; Eph. 4:11; and 1 Pet. 4:10–11.

31. Philosophers of religion concerned with the problem of double agency might find this a useful example.

Judgment, or discernment, in a Charismatic community focuses on distinguishing genuine gifts of the Holy Spirit from acts that resemble them but come from the person alone or from some other source. The criteria for these determinations match very nearly those we have seen above: (1) consistency with Scripture (and, for Catholics, with later church teaching), (2) production of fruits of the Spirit, (3) agreement of the community, and (4) strengthening of the community. The discernment process always begins with the person who feels the urge to speak. The process of learning to recognize impulses from the Holy Spirit is like that Brinton has described in a Quaker worship meeting. Others present are expected to express reactions, positive or negative. The gift might be "confirmed" by the reading of a passage from Scripture, by a simple "Amen!" or by additional prophecies, testimonies, or teachings that support or extend the content of the original message. As an example I quote parts of two "prophecies" given at the 1977 ecumenical conference on Charismatic Renewal in Kansas City. The general point of the first message is confirmed in the second.

> Mourn and weep, for the body of my Son is broken. . . .
> Come before me with sackcloth and ashes, come before me with tears and mourning, for the body of my Son is broken.
> I would have made you a light on a mountaintop, a city glorious and splendorous that all the world would have seen, but the body of my Son is broken.
> Turn from the sins of your fathers. Walk in the ways of my Son.
> Return to the plan of your Father, return to the purpose of your God.

The second message was addressed to church leaders:

> The Lord has a word to speak to the leaders of all the Christian churches. If you are a bishop or a superintendent or a supervisor or an overseer or the head of a Christian movement or organization, this word is for you. The Lord says:
> You are all guilty in my eyes for the condition of my people, who are weak and divided and unprepared. . . .
> This is a hard word, but I want you to hear it. You have not come to me and made important in your lives and in your efforts those things which were most important to me; but instead you chose to put other things first. You have tolerated division among yourselves and grown

used to it. You have not repented for it or fasted for it or sought me to bring it to an end.[32]

Perhaps the most common negative judgment against a putative gift is simply that it evokes no response—it simply fails to affect worshipers as a genuine word from God would be expected to do.[33]

Leadership meetings provide a time for discussion of the long-term effects of members' contributions to the weekly prayer meetings—to note whether or not various supposed gifts and decisions have tended to produce unity and growth in the fruits of the Spirit. Even a decision based on careful discussion is not absolute, however; it is a judgment that the event was probably a gift of the Spirit, and stands pending evidence to the contrary. So long as reasonable precautions have been taken to distinguish between gifts and their counterfeits, a positive judgment entitles members to say of the events that they are acts or words of God.

6. Theological Data: A Proposal

In the preceding sections we have seen that on the basis of the practice of communal discernment, participants in a wide assortment of Christian communities select certain observable events in ordinary church life and designate them as acts (or words) of God. Furthermore, they believe they are entitled to say they *know* that they are acts of God, and explain how they know it by referring to the criteria and practices described above. If these claims to know about God's involvement in certain events appeared only occasionally or in fringe groups, one might be inclined to dismiss them. But against the background of the episodes in Christian history sketched in this chapter, the claims have some warrant to be taken seriously. Such judgment practices, aimed at detecting the work of the Holy Spirit in human life, go not at all beyond what the New Testament says Christian communities are able to do, and they follow the rules offered by some of the most influential of Christian leaders. In

32. Reported by Bruce Yocum, in "He Has Not Spoken in Secret," *New Covenant*, February 1978, p. 6.

33. Cf. Pilgram Marpeck: "All trees are to have fruit, wherever and whenever Christ arrives, be it in season or out of season." "Two Letters," p. 199.

Edwards's work we even see an explanation of why these rules are the appropriate ones for such judgments. He has shown us a plausible theoretical connection between the nature of God (love), God's manner of operation (indwelling of the Holy Spirit), and the effects (making the person like Jesus) so that we understand why appearance of the "fruits of the Spirit" is a defining criterion for the action of God in a person's or a community's life.

Although we have found agreement on the possibility of discernment and great overlap in lists of signs and criteria to be used, there is a diversity in views about the kinds of events and phenomena that are appropriately to be so judged.

During the New Testament era the community was responsible for making judgments regarding the sources of prophecy and teaching; prophets and teachers spoke to both doctrinal and moral or disciplinary issues.

According to Edwards, the pastor judges emotional, moral, and devotional responses to revival preaching.

According to Ignatius, the spiritual director and retreatant discern the presence of the Holy Spirit in order to determine whether a vocational decision was made in accordance with the call of God.

The Anabaptist community evaluates teachings that pertain to doctrine, morality, and discipline.

The Quakers judge contributions to worship and business meetings as to whether or not they are words of God.

Charismatics evaluate putative gifts of the Spirit to determine whether or not they are genuine. Both Quakers and Charismatics assume that the individual speaking will have exercised some prior judgment regarding the source of the impulse to speak.

We might also add to the list Tyrrell's suggestion that the Spirit allows Christians to perceive God's acts in history. Such perceptions would be subject to confirmation by means of a discernment process.

Thus discernment results in judgments of various sorts. It is these judgments that I claim to be an important and often overlooked category of data for theology. The following are some examples of the kinds of facts, relevant to theology, that might result from the discernment process:

> (1) God brought Sarah Edwards to a conviction of his perfect justice.

(2) God called Ignatius to the life of a celibate priest.

(3) God forbids Christians to use the sword.

(4) God commanded the Quakers to free their slaves.

(5) God called Charismatics to repent the disunity among their churches.

I recognize that some Christians might not be comfortable with statements that make such direct claims about God's words and deeds. They might prefer instead to say something like the following: "Sarah Edwards became convinced of God's justice, and Jonathan judged this to be the work of the Holy Spirit." This 'observer's' formulation is certainly the more appropriate one in cases where there is some doubt about the judgment. I claim, however, that if the *theory* advanced by Edwards et al. regarding the possibility of recognizing the acts of God in believers' lives is true, and if there is no good reason to doubt that the discernment process was carried out responsibly, then the more direct 'participant' description is entirely appropriate. Furthermore, it is entirely appropriate for participants in a discernment process to say they *know* that God so convinced Sarah.

6.1 Christian Epistemic Practices

Let us call the practice of making knowledge claims about God's activity in human life on the basis of discernment a *Christian epistemic practice*. To see why such knowledge claims are appropriate we turn to the work of William Alston. "I take as my starting point," says Alston,

> the conviction that somehow what goes on in the experience of leading the Christian life provides some ground for Christian belief, makes some contribution to the rationality of Christian belief. We sometimes feel the presence of God; we get glimpses, at least, of God's will for us; we feel the Holy Spirit at work in our lives, guiding us, strengthening us, enabling us to love other people in a new way; we hear God speaking to us in the Bible, in preaching, or in the words and actions of our fellow Christians. Because of all this we are more justified in our Christian beliefs than we would have been otherwise.[34]

34. "Christian Experience and Christian Belief," in *Faith and Rationality: Reason and Belief in God*, ed. Alvin Plantinga and Nicholas Wolterstorff (Notre Dame: University of Notre Dame Press, 1983), p. 103.

Certain items of Christian belief have more obvious connections with Christian experience than others. For example, the belief that God will reveal his will to those who truly seek him, or that the Spirit is at work in the church, leads to expectations that God will manifest himself in certain ways in individual or corporate experience. When such manifestations occur, they provide empirical confirmation for the beliefs in question.

Alston points out that the most serious challenge to his view is the need to justify conceptualizing experience in theistic terms in the first place. He notes, as I have above, the tendency to bifurcate Christian experience into a psychological datum and a theological explanation. Once reference to God is removed from the description of the experience, it will be very difficult to justify reference to God in the explanation. So the question is how to justify theistic descriptions of experience.

Here Alston turns to a comparison with the epistemology of perception. Philosophers, beginning with Descartes, have pointed out that it is possible to bifurcate perceptual experience into a sense datum ("I seem to be seeing a tree") and a material-object explanation of the experience—a tree to explain the appearance. Once this bifurcation is made, however, direct justification of material-object language may be impossible. Nonetheless, we have a "perceptual practice" of objectifying sense experience. This practice can be justified only in that we have no good reason for supposing it to be unreliable.

Likewise, Alston claims, we have a Christian practice of objectifying certain ranges of experience in terms of Christian theology. By parity of reasoning, we should be justified in this practice so long as there is no good reason for supposing it to be unreliable. Prima facie there are such reasons, however, and they have to do with differences between perceptual practices and Christian practices: (1) There are standard ways of checking the accuracy of perceptual beliefs—by further observation, by intersubjective agreement, and the like. (2) By engaging in perceptual practices, we can discover regularities in the behavior of objects putatively observed and can thereby predict future events. (3) Capacity for perceptual practices is found universally among normal adults. (4) All normal adults, whatever their culture, use basically the same conceptual scheme for objectifying sense experience.

Assuming that these four statements are true of perceptual practices but not true of Christian practices, Alston must show how
Christian practices might nonetheless be reliable. He points out that
on the basis of the following assumptions we should *expect* Christian
practices to lack features (1) through (4): (A) God is too different, too
wholly other, for us to grasp any regularities in his behavior. (B) For
the same reason we can attain only the faintest, sketchiest grasp of
what God is like. (C) God has decreed that a human being will be
aware of his presence only under special circumstances.

Alston notes that (A) through (C) are in fact standard parts of
Christian teaching, and are compatible with three further assumptions: (D) that religious experience should, in general, constitute
genuine awareness of the divine; (E) that although any particular
articulation of such an experience might be mistaken to a greater or
lesser extent, indeed even though all such articulations might miss
the mark to some extent, still such judgments will, for the most part,
contain some measure of truth; and (F) that God's designs contain
provision for correction and refinement, for increasing the accuracy
of the beliefs derived from religious experience. We have no reason
for supposing that (D) through (F) do not obtain, and therefore
Christian practice is justified. Further justification for thinking
Christian practice is reliable is the fact that the Christian enterprise as
a whole bears its own proper fruit in the transformation of lives.

Now, the skeptic reading Alston's argument might quickly object
that his invocation of hypotheses about the nature and activity of
God begs the question. For this reason I believe it is more helpful to
turn to analogies with science rather than the epistemology of perception. If Alston could show (A) through (F) to be auxiliary hypotheses in a progressive theological research program then, indeed, there
would be an element of circularity in using data to support theories
that in turn justify the very use of those data. But this circularity is
characteristic of all research programs. The question then is whether
the research program as a whole is acceptable, and we have Lakatos's
criterion of empirical progress to answer that question.

My suggestion is that in place of (A) through (F) an amended
Edwardsean theory of discernment be added as an auxiliary hypothesis to all relevant theological research programs. Here we have a
theory about how God's actions are to be recognized, criteria to use
for such judgments, and an account of the nature and operation of

God that explains why these criteria are the relevant ones. This theory readily accounts for statements very close to Alston's (D) through (F). First, it recognizes *some* 'religious experience' as genuine awareness of and response to the divine. Second, the communal element with which we have supplemented Edwards's account is a direct response to the recognition that individuals may make mistaken judgments. Even the judgments of the community are tentative and stand only until disconfirming factors are brought to light. Finally, for Alston an epistemic practice is apparently a simple matter of describing certain kinds of experiences in certain kinds of terms.[35] I use the term to refer to the practice of communal discernment and of describing events so judged in 'participant' language. Furthermore, the practice I envision admits of development, as does a MacIntyrian practice. Alasdair MacIntyre defines a practice as "any coherent and complex form of socially established cooperative human activity through which goods internal to that form of activity are realized in the course of trying to achieve those standards of excellence which are appropriate to, and partially definitive of, that form of activity, with the result that human powers to achieve excellence, and human conceptions of the ends and goods involved, are systematically extended."[36] Christians may begin with very inadequate notions of how to listen for God's voice and with poor skills in discernment. But practice in discerning produces a community with a much more sophisticated view of God and God's workings, and thus with greater discernment. Critics may complain that the discernment criteria that emerge from combining the accounts of Edwards and the others are too vague to be useful; for instance, there may be too much disagreement over what it means to display a Christlike character. This criticism fails to take account of the self-correcting character of a discerning community. If the good to be achieved by discernment is the development of a Christlike community, and if discernment is a practice in MacIntyre's sense, then we may expect the community's understanding of the good to develop along with skills involved in the practice. So here we have an account

35. See his forthcoming *Perceiving God,* however, for a more adequate account.
36. *After Virtue,* 2d ed. (Notre Dame, Ind.: University of Notre Dame Press, 1984), p. 187.

of how God's provision for correction and refinement (Alston's hypothesis F) is made concrete.

In his account of Christian epistemic practices Alston assumed the most difficult case: no predictability in the realm of Christian experience; no regularities to be observed in God's actions; therefore, no direct way to establish the reliability of Christian epistemic practices. A further advantage of locating the relevant practices at the level I have chosen is that here measures of reliability are possible. We take up this issue in Section 6.3 below.

6.2 Theory-laden Data

I have suggested that judgments resulting from Christian discernment constitute suitable data for theology. The discernment process results in many cases in direct claims to knowledge about the activity of God (e.g., "God commanded the Quakers to free their slaves"). I noted the tendency many theologians (and other Christians) might have to make more guarded statements (e.g., "The Quakers freed their slaves because they judged that this was God's command"). I then turned to William Alston for aid in supporting my claim that the participants' description of the results of discernment is at least as appropriate as the more guarded claim. Alston argues that description of religious experience in theistic terms is an acceptable epistemic practice. I would argue that with the confirmation provided by *communal* discernment, and in light of a theory such as Edwards's, we have even more reason to describe certain phenomena in church life as acts or words of God.

The point of arguing for participant descriptions of the results of discernment is that such descriptions can then provide data for theology *that already contain reference to God*. To appreciate the significance of such data we must take another detour through philosophy of science.

In 1958 N. R. Hanson published *Patterns of Discovery*, one of whose major theses has now become widely accepted by philosophers of science, namely, that there is no pure, uninterpreted observation but, rather, all facts are "theory laden."[37] Hanson believed that philosophers erred in their discussions of perception by taking patches of

37. Cambridge: Cambridge University Press, 1958.

color and the like as their paradigm cases of seeing. It would be better to regard as paradigms seeing what time it is, seeing what key a piece of music is in, or seeing whether a wound is septic. Seeing itself is influenced by what one knows and by the language used to express that knowledge.

According to Hanson, 'seeing that . . .' is the logical element that connects observations with knowledge and with language. To see boxes, birds, rabbits, goblets, faces, x-ray tubes, is to have knowledge. One very important type of knowledge already present in seeing is knowledge of causes; for example, seeing a concavity as a crater is already to be committed as to its origin. Without the knowledge already contained in 'seeing that . . .' or 'seeing as . . .', one would never be able to reason from observations to their explanations. Theories provide patterns within which data appear intelligible. A theory is not pieced together from observed phenomena; it is rather what makes it possible to observe phenomena as being of a certain sort and as related to other phenomena. This is not to say, of course, that theories create what is seen, only that theoretical knowledge allows the observer to organize the raw data of sensation into intelligible patterns. It does leave open the possibility, however, that there may be more than one intelligible pattern.

In order to apply Hanson's discussion to the question of data for theology, focusing specifically on the results of discernment practices, we need only change 'seeing that . . .' to 'perceiving that . . .' to take account of the fact that observation of God's acts and hearing God's word involve various forms of perception other than vision. As in science and, in fact, knowledge generally, we have here *theory-laden facts*. In the clearest cases Christians do not say that they heard the words of fellow believers and then interpreted them as God's; rather they hear God speaking through the human speaker; the community's discussion or response attempts simply to find whether others heard it as well. The experience comes interpreted, but this is no objection since that is the regular means by which observation becomes knowledge. The surest way to get from observations to hypotheses or theories that explain them is to begin with observations that are already expressed in language suggestive of the causes or of the explanatory framework. The value for theology of observations already communally described as acts of God is obvious.[38] In

38. Hanson's account explains the fact that evidence for creation from the order of

short, if God does not appear in the facts, his presence in the explanation will always be suspect.[39]

6.3 The Objectivity of Theological Data

I have claimed in the preceding pages that the results of Christian discernment might well serve as data for theology, despite the fact that such facts involve interpretation. But if theory-ladenness is no objection, there might well be others. One might object that scientific data are objective, whereas these observations are subjective. Now, 'objectivity' is used in a number of ways. The most appropriate use for our purposes might be that of social scientists, whose difficulties in measurement are almost as great as those of the empirical theologian. When the social scientist makes a measurement (which may be a nominal measurement—simply determining whether some property or state of affairs obtains or not) objectivity depends on two variables: validity and reliability.

If a measure is reliable, then we expect to get roughly or exactly the same results under similar circumstances. Reliability is always a matter of degree; different degrees are required depending upon the complexity of the matter under study. Measurement with a ruler is highly reliable; measurement with an I.Q. test is only moderately reliable. We have no data on the reliability of believers' judgments regarding the presence or absence of God's agency in certain events. Whether or not a particular community's judgments are as reliable as those accepted by scientists for equally complex material is open to investigation. But communities that exercise communal judgment do not readily abandon the practice. This fact suggests that reaching agreement is possible (even if tedious in some cases) and that results tend to be consistent over time. A practice that yielded erratic results would soon lose its appeal.

---

nature seems so conclusive to those who are already convinced theists and so worthless to those who are not: the theist already *sees* the world as a creature; the atheist does not.

39. Wayne Proudfoot, on the other hand, claims that "religious experience" must be *described* in terms acceptable to the experiencer, but may then be *explained* in entirely different terms—for example, by means of psychological theories. I believe he has not taken adequate account of the way description 'stacks the deck' in favor of a particular explanation. See *Religious Experience* (Berkeley: University of California Press, 1985).

One might object that the variety of beliefs and practices found throughout the Christian movement across denominations and through time argues for the unreliability of discernment. I suggest, however, that it argues instead for the need to make more frequent and determined use of the practice. In fact, much of what goes in on in the life of many churches has never been subjected to prayerful testing. A positive outcome of the acceptance by theologians of my proposal to seek data for theology among the results of discernment might be an increased interest in the procedure within the churches. If such were the case, the church would become the locus for testing theological formulations—laboratories for theological experimentation.

The validity of a measurement refers to its genuine connection with the thing measured—for example, is academic ability a valid measure of intelligence? The value of Edwards's theology, as I suggested above, is that it provides a rationale for expecting the signs of gracious affections to be true indicators of God's activity, not because of some accidental connection, but because they participate in the divine character they represent. If Edwards is correct, then the criteria by which Christians judge the genuineness of impulses from the Holy Spirit do provide a *valid* measure of God's involvement in assorted events.

A second use of 'objectivity' in a consideration of scientific observations is to indicate that they are open to public observation rather than private. The proposed theological facts are objective in this sense as well. Discernment is a process for judging the nature of certain events as to whether they are God-inspired or not. The events themselves are public: people speaking, or praying, or undertaking assorted activities. The privacy, if there is any, is in the community member's perception that the event is God-inspired. But notice that the same is true in cases of 'seeing that . . .' in any of the sciences. Seeing that the battery is charged, or that the titration is complete, or that the rat has learned the maze, or that the subject's ego strength has increased all involve the scientist's own private perceptual-plus-interpretive experience. 'Objectivity' just means that others under similar circumstances and with similar theoretical training and experience will see the same thing, and this again is a matter of degree. Here we see more fully the point of *communal* discernment. The question is raised whether all those present perceived the same thing, or whether one person has misread the situation. If there is no immediate agreement, further tests can be applied as the event's consequences unfold.

6.4 Novelty versus Replicability

The theologian working by Lakatosian standards is in a difficult position in seeking suitable data for theology. On the one hand, critics may claim that none of theology's data are acceptable because they are unique events—not replicable as is a laboratory experiment. On the other hand, the demands of methodological sophistication call for novel facts. Fortunately, though, the epistemic practice described above can satisfy both requirements. To see this, we have only to distinguish between the judgment itself and the procedure by which the judgment is made. The judgment may indeed be novel relative to a particular theological research program, while the discernment procedure is repeatable. Now, most conclusions reached by a discerning church will not be of a new sort. The worship and moral life of the Christian church evolve slowly, and the criteria for discernment have a strongly conservative tendency; the requirement of consistency with the apostolic witness makes radical change rather unlikely. Nonetheless, the church must be open to the possibility that God will do new things in its midst and to the possibility that previous generations have misinterpreted the first witnesses. Recall the difference between Edwards and the Anabaptists in the extent to which they found the Spirit's work a creative force in the fomulation of doctrine and practice.

To see how a novel yet replicable fact might be found to support a theological research program, we turn to the writings of Jesuit theologian Donald Gelpi. In his *Divine Mother: A Trinitarian Theology of the Holy Spirit* Gelpi shows that there is precedent for employing feminine images to represent the Spirit and recommends using feminine pronouns to refer to "her."[40] He also provides a complex theoretical rationale for the appropriateness of this new linguistic practice. If Gelpi's historical and constructive work is sound, then his theory predicts that prayer addressed to the Holy Spirit as "she" will become an accepted practice. An important concern, of course, is the

40. Lanham, Md.: University Press of America, 1984.

force of 'accepted practice.' If it were simply imposed on worshipers for the sake of a theological theory, it would be useless as evidence for that theory. It would be a significant fact for theology only if it passed the test of the *discernment* of a number of worshiping communities. If it were widely accepted as evidently in accord with the promptings of the Holy Spirit (herself), then that judgment would provide important information about the very nature of God and it would indeed be a *new* fact, established by means of a *replicable* procedure.

I defined a novel fact (in Chapter 3) as one not used in the construction of a theory T that it is taken to confirm. A fact not used in the construction of T is one whose existence, relevance to T, or interpretability in light of T is first documented after T is proposed. By hypothesis, the widespread practice of worshiping a feminized Spirit, as well as the associated judgment that the practice reflects the will of God, is a fact that comes into existence only after the doctrinal theories developed in *The Divine Mother* were added to Gelpi's theological research program, and would therefore be *novel* with respect to that program.

The discernment process that establishes the appropriateness of such worship is a *replicable* process. Just as scientists can repeat experiments to check low-level generalizations, so too can Christian communities repeat the judgment process that leads to the conclusion that such worship of the Spirit is in accord with the will of God. Not every attempt at discernment produces clear results, of course, but neither does every attempt to replicate an experiment, especially in the human sciences. Therefore I conclude that theological facts of this sort are often or ordinarily replicable and thus do not differ from scientific facts in this regard.

7. Scripture as Data for Theology

No account of theology would be complete without considering Scripture. Is not most Christian theology supported by Scripture study of some sort, and if so, do the texts not constitute data for theology?

A Lakatosian approach suggests that the role of Scripture in theological research programs cannot be legislated beforehand, but must rather be governed by auxiliary hypotheses built into the various programs. David Kelsey has made a comparable proposal in his study titled *The Uses of Scripture in Recent Theology*.[41]

Kelsey points out that 'Scripture is authoritative for theology' is analytic: Calling a text 'Scripture' *means* that it is authoritative for the community that claims it as such; it is to be used in certain ways in the common life of the community in order to establish and preserve the community's identity. The texts' authority for theology is logically grounded in and dependent on their authority for the life of the church generally, since theology is the activity in which "forms of speech and action current in the church are self-consciously criticized as to their aptness or faithfulness, and in which proposals are made for reform of those forms of speech and action so that they will be more faithful."[42] Part of the community's faithfulness must be faithfulness to its Scriptures. Kelsey concluded from the examples he selected (seven Protestant theologians writing mainly between 1920 and 1960) that to say that Christian Scripture is authoritative for theology can be only a formal statement, however, because Scripture functions differently in each case. The way it is used depends on the theologian's logically prior judgment about how best to construe the manner of God's presence in the community (ideationally, as agent in history, as ideal possibility, or in some other way). The theologian attempts to "catch up what Christianity is basically all about in a single, synoptic, imaginative judgment."[43] This imaginative construal is in each case grounded in features of the common life of the church.

Kelsey's view of the role of Scripture in theology is quite compatible with a research-programs approach. His single, synoptic, imaginative judgment invites comparison with Lakatos's notion of a hard core. In both cases, the central idea logically precedes particular decisions about how to use Scripture. In line with Kelsey, I offer here no general theory of the role of Scripture in theology, but will instead review some of the ways Scripture has been involved in research programs so far presented.

For Tyrrell, Scripture serves as a record of religious experiences of

41. Philadelphia: Fortress Press, 1975.
42. *Uses of Scripture*, p. 99.
43. Ibid., p. 159.

various sorts, expressed in poetic, imaginative language, which theology attempts to unify and systematize in conceptual language. An important part of the content of Scripture is "prophetic history," that is, history told from God's point of view, making his intentions apparent. Thus the texts include a kind of truth inaccessible to ordinary historical inquiry.

For both Pannenberg and the Lundensian theologians (to be considered in the following chapter), Scriptures are historical texts. According to Pannenberg, the theologian must use the texts as would any scientific historian. For the Lundensians, theology is a discipline that is sui generis—it seeks a peculiar religious content in the texts, a content that is not the province of any other discipline.

Earlier in this chapter we have also seen a variety of ways in which scriptural texts function in the life of the community. For both the Anabaptists and Edwards, Scripture functions as a source of criteria for recognizing the presence and action of God in the lives of the faithful. For the Charismatics it functions as well as a source of examples of the operation (gifts) of the Holy Spirit. And most important, for Anabaptists, Charismatics, Edwards, and Ignatius, it functions as a general guide to Christian practice.

So we have found just what Kelsey's functional view of Scripture would lead us to expect: the Scriptures are authoritative for all the individuals and groups surveyed, but exactly how they function varies considerably, from records of religious experience, to historical documents, to norms for community life.

In terms of the present work, we may say that the use of Scripture in theological research programs is governed by a variety of *auxiliary hypotheses*—usually doctrines of revelation. Pannenberg, for example, has elaborated a theory of revelation as history; Tyrrell has a theory of revelation through religious experience. Contemporary Charismatics hold a theory of revelation through the gifts of the Spirit, assuming that the texts were written with the aid of the same kinds of impulses experienced today as gifts of the Spirit, and that the authority of the text rests on the discernment of the early church—that is, on its communal judgment that genuine words and acts of God are recorded therein.[44] In scientific terms, one might say that

44. See Richard Quebedeaux, *The New Charismatics II* (San Francisco: Harper and Row, 1983), pp. 133, 135.

contemporary experience of the gifts serves as a model for a theory of revelation.

The Charismatics' understanding of Scripture is quite interesting from the point of view of this book because of its consistency with the theory of discernment advocated here. In this case, a single theory regarding the manner of the Spirit's operation legitimates the use as data of both Scriptural texts and judgments regarding ongoing events in the life of the church.

Although it is common to employ a doctrine of revelation to explain and justify a community's use of Scripture, it is not necessary to do so. James McClendon explains the particular way churches in the Anabaptist tradition use Scripture by means of a doctrine of the church, which 'identifies' the church today with that of the New Testament era and calls its members to the same discipleship as those who followed the earthly Jesus. This "baptist vision" entails that the Scriptures be read as addressed with the same force to the church today as to the communities for which they were originally written.[45]

So we see a variety of doctrines (auxiliary hypotheses) proposed to explain the authoritative role of Scripture in theology and in the life of the church generally. A doctrine that explains well is thereby confirmed. If the doctrine forms a part of a systematic theological research program, however, its final justification is a function of the empirical progress (in Lakatos's sense) of the entire research program in comparison with its competitors.

Beyond the question of authority (but not unrelated to it) is another complication. When we attempt to look at Scripture as a source of data for a scientific theology, we see that in effect there is a hierarchy of levels of data. At the lowest level we find ancient autographs and other sources that go into the production of a critical text, which in turn provides the data for the translator and the biblical scholar. The biblical scholar's conclusions (theories) become data for the theologian. Theoretical assumptions of various sorts are required for justification of the conclusions at each of the higher levels. The theologian's theories about the nature of Scripture and revelation will, in dialectical manner, partially determine which of these levels will be seen as most relevant and how the conclusions will be taken.

45. James William McClendon, Jr., *Ethics* (Nashville: Abingdon, 1986), chap. 1.

Notice that if it is the conclusions of biblical scholars that provide the most immediate level of data for theology, there is no reason not to expect *occasional* novel facts in this area. It is possible to predict, on the basis of a new modification in a theological research program, that biblical scholars will find something specific in the Scriptures, and that prediction may be borne out in the course of further research. The next chapter will provide an example of novel corroboration of a theological research program on the basis of new discoveries by biblical scholars.

8. Conclusion

The problem I have attempted to address in this chapter is that of finding suitable data for theology. I have briefly discussed data from Scripture and have suggested that occasional novel facts might be found among the results of biblical criticism. But the requirement for novel facts is more likely to be met from within a domain of ongoing events. I proposed as a candidate the results of Christian discernment: judgments regarding the involvement of God in assorted events in the life of the church. Facts of this sort are likely to be particularly useful for theology for two reasons. First, these are facts *about God*. When no mention of God appears in the data, it is difficult to justify the invoking of God in the explanation. Second, because the surest means of coming to know a person is through words and deeds, the most relevant among facts about a personal God would seem to be those reporting God's words and actions.

As I mentioned above, the philosopher is in no position to dictate what may or may not be found relevant to support a scientific *or* theological theory. Also, relevant facts will not all fall into one neat category (consider the variety of kinds of facts that might be relevant in supporting a psychological theory: human speech and behavior; results of psychological tests; physiological findings; animal behavior). But the Lakatosian philosopher *is* in a position to demand that any research program contain auxiliary hypotheses to validate the kind or kinds of data it uses. That is, it must provide some rationale for the supposition that the facts in question are the sort of thing that will yield knowledge about the entities studied (in this case, God). I have claimed that Edwards's theory of the religious affections pro-

vides a good model for such a rationale, justifying the use of judgments from discernment as data for theology.

I have also claimed that despite possible arguments to the contrary,
the judgments to which I have called attention meet all the standard
requirements for scientific data. They will not be of the same quality
(reliability, replicability) as those of the natural sciences; they may
more justly be compared to those of the human sciences such as
psychology.[46] Furthermore, discernment being a practice available
to any group of Christians, theological experimentation is not at all
impossible.

46. My conclusions here are consistent with Ian Barbour's in *Myths, Models and Paradigms* (New York: Harper and Row, 1974).

[6]

Scientific Theology

> Those who have a propensity to philosophy will . . . con-
> tinue their researches, because they reflect that, besides the
> immediate pleasure attending such an occupation, philosoph-
> ical decisions are nothing but the reflections of common life,
> methodized and corrected. But they will never be tempted to
> go beyond common life so long as they consider the imper-
> fection of those faculties which they employ, their narrow
> reach, and their inaccurate operations.
>
> —DAVID HUME, *An Enquiry concerning
> Human Understanding*

So far I have raised the problem of the cognitive status of theology:
Can it make claims to knowledge of God in the age of probable
reasoning? I pointed out (Chapter 1) that there is at least a minority
tradition in modern theology that answers in the affirmative—Wolfhart Pannenberg being one notable instance. I took issue with Pannenberg's account of the standards of probable reasoning, however,
on the grounds that his methodology is in the end unworkable. I then
presented (Chapter 3) a different view of scientific rationality. Imre
Lakatos is but one of the more important recent philosophers of
science, but his methodology goes further than any other toward
giving a *workable* criterion for choosing among competing scientific
research programs. If theological programs could be shown to be like
scientific research programs, Lakatos's criterion of relative empirical
progress could be used to adjudicate between competing theological
schools as well. The more acceptable program(s) could then claim to
provide knowledge of God and God's relation to the world.

The next step in my argument was to examine a theological program (Chapter 4) and show that in fact it has all the features of a

scientific research program: hard core, expanding belt of auxiliary hypotheses, data, and positive heuristic. What appeared to be lacking in the modernist program was a clear sense of how to isolate data that were independent enough of the theoretical structure to count as corroboration of it. So I supplied (Chapter 5) an account of theological data as judgments resulting from Christian discernment. Such judgments reflect the involvement of God in the church's affairs: in its devotional and moral life, in the shaping of its convictions, and perhaps even in its perceptions of history. I also proposed Jonathan Edwards's theory of religious affections as an appropriate auxiliary hypothesis to explain why the signs used in discernment may be taken as defining criteria of the presence and action of God. These data are obviously theory-laden, but no more so than the data in other sciences. Furthermore, discernment is an ongoing process (in some church bodies), so there are opportunities to find new facts that may provide novel confirmation for theological research programs.

There is now but a short step from description to prescription— from describing the sciencelike character of (some) theology to advocating a deliberate adoption of Lakatos's methodology in theology. This chapter will spell out in more detail what this step would mean, and then relate this proposal to other recent developments in philosophy of religion, philosophical theology, and work on the relations between theology and science.

1. Scientific Doctrinal Theology

My example in Chapter 4 of the application of Lakatos's theory of research programs to theology is not as helpful as it might be in illustrating how the main business of theology is to be carried on in its light. The modernist program was a program in fundamental, or foundational, or philosophical theology—of the same genre as the present work, while doctrinal theology (theology proper) involves the Christian theologian in reflection on a fairly standard list of topics or loci: the nature of God and the Trinity, Christology, sin and soteriology, atonement, pneumatology, ecclesiology, and so on. Some theologians address themselves to a single doctrine; others attempt a comprehensive treatment. A single doctrine could form the center of a theological research program, but the research-programs model seems to lend itself better to incarnation in a systematic theology involving many doctrines. Here some central organizing idea would serve as the hard core of the program, while theories regarding the various doctrines would constitute auxiliary hypotheses elaborating that central idea and relating it to the data. Additional auxiliary hypotheses would provide a rationale for taking those facts as relevant support for the program and would describe the program's concrete methods of research.[1] To clarify this proposal I present two examples of doctrinal theologies construed as research programs along the lines I have just suggested.

1.1 Pannenberg's Theology as a Research Program

Let us first recast Pannenberg's theology, as explained in Chapter 2, in the form of a doctrinal research program. Here the hard core would be his claim that the God of Jesus is the all-determining reality. This is the central and quite abstract theory that guides the development of the entire program.

The positive heuristic of any systematic theology, I believe, is the plan to develop theories (auxiliary hypotheses) concerning all the traditional theological loci in such a way that they meet the following conditions: (1) they are faithful to any authoritative pronouncements within the relevant communion; and (2) they serve to elaborate or spell out the content of the hard core in a way that (3) relates the doctrines to available data.[2] In Pannenberg's case, the nature of the hard core—the insistence on God's relation to all that is—requires that the data include not only the scriptural texts, but facts and theories from all areas of knowledge.

Pannenberg's theory of revelation as history, along with more specific theories of interpretation (for example, the thesis that the principle of analogy is to be used only in a qualified way in the interpretation of historical texts) constitute important methodologi-

---

1. For somewhat different uses of Lakatos's methodology for understanding theology, see Philip Hefner, "Theology's Truth and Scientific Formulation," *Zygon* 23, no. 3 (1988), pp. 263–79; Hefner, "The Role of Science in Pannenberg's Theological Thinking," in *The Theology of Wolfhart Pannenberg: Twelve American Critiques*, ed. Philip Clayton and Carl Braaten (Minneapolis: Augsburg, 1988), pp. 266–86; and Philip Clayton, *Explanation from Physics to Theology: An Essay in Rationality and Religion* (New Haven: Yale University Press, 1989).

2. For a more detailed discussion of the positive heuristic, see below, sec. 1.3.3.

cal auxiliary hypotheses comparable to theories of observation or instrumentation in science.

A crucial line of argument in support of the hard core runs from the Pauline texts regarding the appearances of Jesus after his burial along with other historical data (such as the lack of any accounts of recovering his body), through a cluster of auxiliary hypotheses regarding Jesus' resurrection from the dead and its significance against the background of Jewish apocalyptic, to the conclusion that God is definitively revealed in Jesus. This conclusion has implications both for Pannenberg's doctrine of God (the hard core) and for his Christology (an important cluster of doctrinal auxiliary hypotheses).

Additional clusters of doctrinal hypotheses would have to include other traditional Christian doctrines: ecclesiology, Spirit, Trinity, Christian anthropology, doctrine of creation, and so on.

Let us return now to the notion of the positive heuristic. Pannenberg's hard core, which refers to God not only as the God of Jesus Christ but also as the all-determining reality, adds an extra burden to his positive heuristic. Besides reinterpretation of the tradition, his program calls for confirmation of the hard core by means of data from all branches of knowledge. Thus part of his plan is to develop auxiliary hypotheses to mediate (form logical links) between such extratheological reality and the core. Ideally these additional auxiliary hypotheses will not be unrelated to the doctrinal auxiliaries. Rather, the doctrines themselves will mediate between the core and both scriptural and experiential data. This is in fact the kind of development we see in Pannenberg's later theology. Let us consider in particular the development we see in his recent *Anthropology in Theological Perspective.*[3]

Systematic theology traditionally has had an interest in a Christian anthropology. Pannenberg develops theories in this area that serve as auxiliary hypotheses to account for the data of the Christian tradition and for particular facts of human experience as noted by secular anthropologists: the human phenomena he describes as exocentricity, ecstasy, and spirit. But Pannenberg does not simply take over the works of anthropologists; rather, he offers reinterpretations of their

---

3. Philadelphia: Westminster Press, 1985, trans. by Matthew J. O'Connell of *Anthropologie in theologischer Perspektive* (Göttingen: Vandenhoeck und Ruprecht, 1983).

conclusions. Thus it would be more accurate to say that Pannenberg's Christian anthropology explains the data of the Christian
tradition while providing a Christian interpretation of anthropological theories.

Notice that the revision of auxiliary hypotheses in order to create
consistency between data and the hard core is exactly in accord with
Lakatos's description of scientific practice. Lakatos's methodology
requires, however, that such alterations be progressive—that is, that
the new version of the theory allow for prediction of some new fact.
It is not to be expected that each such theoretical modification will
lead to the corroboration of a new fact, but research programs are to
be judged on the basis of the frequency of such occurrences. So the
prediction and corroboration of some fact previously unanticipated
by the anthropologists at this point would go a long way toward
establishing the scientific respectability of Pannenberg's theology.

Thus Pannenberg's system, when viewed as a theological research
program (and if we ignore his own criteria for scientific success),
looks quite good. It certainly has all the structural features of a
research program; all that is lacking is to determine whether it is
empirically, as well as theoretically, progressive.[4]

1.2 The Lundensian Program

For contrast, let us look at another quite different Lutheran program. I also include this example because recent biblical scholarship
has turned up novel corroboration for the program, some years after
it had slipped into obscurity—illustrating Lakatos's point that a
degenerating program can sometimes be rescued by new evidence.
This illustration comes from the school of theology dominant at the
University of Lund (Sweden) from the early 1920s to the late 1930s.

---

4. At a symposium on his work at the Lutheran School of Theology at Chicago
(November 15–17, 1988), Pannenberg accepted my reconstruction of his program
and accordingly indicated several predictions whose corroboration would offer relevant confirmation: for example, that any anthropology that turns out to be successful
scientifically will have to provide a more constructive place for religion than those of
the recent past, and that physics will develop a new form of field theory that will
include the irreversibility of time. Note that these predictions are of a theoretical
nature, but that is appropriate for a research program whose theories are of a very
high level.

The architects of the system were Anders Nygren, Gustaf Aulén, and Ragnar Bring.

In his history of the movement, Nels S. F. Ferré characterized Lundensian theology as a "resurgence of historic Christianity," "a new Biblicism," recognizing Luther as a final authority on Scripture.[5] In fact, though, the starting point for this school was philosophical. Just as Kant had "transcendentally deduced" the categories and the pure intuitions of space and time so as to validate their contents, Nygren set out to 'deduce' a religious form of consciousness.[6] He reasoned that if there were a religious form of consciousness, then there must be a religious content accessible to historical investigation. Theology's task would then be to find and describe this content. Christian theology describes the content of Christian consciousness; Jewish, or Hindu, or Buddhist 'theology' describes Jewish, Hindu, and Buddhist forms. Religions are not mere multitudes of facts, however; observation shows that their ideas are unified by basic 'motifs.'

In Nygren's "transcendental deduction" of the religious category he intended to show that this form of consciousness is necessary and transsubjectively valid. He proceeded as follows: The judgments of science, aesthetics, and ethics all claim validity. But a judgment cannot be valid only for a time, so the ascription of validity to a judgment presupposes the category of eternity. The category of eternity is the basic category of religion, which is 'filled in' by the content of the various religions. Thus the religious *form* of experience is necessary and transsubjectively valid, because without it there would be no basis for any judgments about experience, even though the *content* of that category is variable and contingent.[7]

The doctrinal theology of the Lundensian school focused on the motif of *agape*, God's transcendent, unmotivated love, and displayed its significance in Christian thought by relating it to other theological *loci*.[8] The best-known example of this in English is Aulén's explica-

---

5. *Swedish Contributions to Modern Theology* (New York: Harper and Row, 1967), pp. 23, 25, 29.

6. In his *Religiöst apriori, dess filosofiska förusättningar och teologiska konsekvenser* (Lund: Gleerup, 1921).

7. Ferré, *Swedish Contributions*, pp. 41–46.

8. See Nygren's *Agape and Eros: A Study of the Christian Idea of Love*, authorized trans. A. G. Hebert (London: S.P.C.K., 1932).

tion of the "classical" doctrine of atonement.[9] The Lundensians' method required them to show each doctrinal position in Lutheran theology to be a consequence of the understanding of God represented by the *agape* motif, while the individual doctrines themselves were to be supported by the fact that they served as unifying motifs for significant parts of the history of Christian thought.

I believe that the hard core of the Lundensian program can be expressed as follows:

> The content of the (transcendentally valid) category of the eternal in Christian theology is faith in the God of pure *agape*, who is definitively revealed in the Christ-deed.

The philosophical basis of the Lundensian program—the transcendental deduction of the category of the eternal—may be read as a complex auxiliary hypothesis explaining what is meant by the phrase "category of the eternal" in the hard core. Notice that a transcendental deduction is actually an instance of hypothetico-deductive reasoning where only one hypothesis is imaginable. One begins with a known fact—in this case, the 'fact' that there are valid judgments. Then one asks what must be the case in order for this fact to be possible. So (if we reduce Nygren's argument to a single sentence) the statement: "there must be a category of the eternal if there are valid judgments" is a mediating premise connecting the core theory to the 'fact' of valid judgments. At the same time, it helps explain the meaning of the phrase "category of the eternal" as it occurs in the hard core. This is the normal function of an auxiliary hypothesis. Methodological auxiliary hypotheses for the program derive from this philosophical basis; they include directions to seek the content in the history of religions and to organize it according to motifs.

The auxiliary hypotheses of a doctrinal nature begin with a doctrine of God based on analysis of the concept of *agape*, with emphases on God's sovereignty and on the consequence that there can be no symmetrical relations between God and humankind. Consequently, revelation is the only possible source of knowledge of God. The doctrine of revelation, in turn, along with specification (in the hard core) that revelation is through the Christ-deed, has consequences for still another methodological auxiliary hypothesis, namely, that a

9. *Christus Victor*, trans. A. G. Hebert (New York: Macmillan, 1931).

major part of the data for the Lundensian research program must be the life of Christ as witnessed in Scripture. Another consequence of the stipulation that there can be no symmetrical relations with God is that, of the three main atonement motifs Aulén found in Christian history, only the classical (ransom) theory was acceptable, because neither the subjective (moral influence) theory nor the Latin (Anselmian) theory shows adequate appreciation for the unidirectional force of God's work in Christ. Further consequences regarding anthropology, Christology, doctrine of sin, eschatology and other doctrines, were spelled out by the Lundensians, related to the hard core and, in Ferré's words, "buttressed by volumes of biblical commentaries to support the structure."[10]

Since its heyday the Lundensian program has slipped into obscurity. Nonetheless, some novel support has come to light for Aulén's atonement theory. Aulén claimed to find three dominant motifs in Christian thought regarding the atonement. Oldest of these is the "classic" idea of divine conflict and victory: Christ fights against and triumphs over the evil powers, who hold humankind in bondage. An early version of this theory is found in the writings of Irenaeus (125?–202), but in time there came to be at least three versions expressing the transaction with the devil: a quasi-legal version elaborating the idea, already in Irenaeus, of a ransom paid to the devil; a second version wherein the devil was deceived; and finally what we may call a political version, according to which the devil lost his dominion through misusing his rights over Christ. For example, Augustine's account may be summarized as follows: The devil found Christ innocent but took him nonetheless; he shed innocent blood and took what he had no right to take. Therefore it was fitting that he should be dethroned and forced to give up those who were under his power.[11]

By Anselm's day (1033?–1109) the classical theory had fallen into disrepute; the demon language seemed crude and mythological. The "Latin" motif, first instantiated in Anselm's *Cur Deus Homo*, interprets the work of Christ in terms of a legal relationship: Christ by his passion and death earned an excess of merit, which is paid to God as satisfaction for sin.

The earliest instantiation of the "subjective" motif, nearly contem-

10. *Swedish Contributions*, p. viii.
11. *Christus Victor*, p. 51.

porary with Anselm's, was proposed by Abelard (1079–1142). Here Christ's life, death, and resurrection effect reconciliation with God by manifesting God's love so powerfully as to inspire human love in return.

By Aulén's account, the Latin and subjective theories, competing since the Middle Ages, have each so clearly brought out the other's weaknesses that neither is acceptable; thus the classic motif is left without a rival. Aulén claims that the unacceptable mythical language *can simply be dropped*, leaving the rest of the doctrine intact. It is not, however, at all clear how to interpret this latter claim (so far) as anything but an ad hoc qualification.

In the 1950s and '60s, New Testament research by G. B. Caird and Hendrik Berkhof brought about a new understanding of the "principalities and powers" in Paul's writings.[12] Although there are lingering connections with evil spiritual beings (the gods of other nations), Paul seems to use these terms to refer to sociopolitical powers: human traditions, the state, class and economic structures, and religious institutions, to name a few. John Howard Yoder has shown that, in light of this fresh understanding, a new instance of the classical atonement theory in its "political" version can now be developed—that is, an account of the sense in which the evil powers lost their rights over humankind by abusing the innocent Christ.[13] The powers are sociopolitical structures that have become idols— they require individuals to serve them as though they are of absolute value. Christ destroyed their idolatrous claims in both his life and death: First, he showed it was possible to live a genuinely free life in spite of the powers. Second, the most worthy representatives of Jewish religion and Roman politics revealed their true colors in putting him to death. Christ disarmed the principalities and powers by stripping them of their ability to create an illusion of absolute legitimacy; he made a public spectacle of them and thereby triumphed over them (cf. Col. 2:15). Insofar as these powers enforce sinful ways of life, Christians are now freed from bondage to sin.[14]

12. G. B. Caird, *Principalities and Powers* (Oxford: Clarendon Press, 1956); Hendrik Berkhof, *Christ and the Powers* (Scottdale, Pa.: Herald Press, 1962).

13. *The Politics of Jesus: Vicit Agnus Noster* (Grand Rapids, Mich.: Eerdmans, 1972), chap. 8.

14. The strength of this theory is that it provides such a straightforward account of how Christians are freed from bondage to (some) sin—that is, from bondage to

Thus the demonology of the classic atonement theory can be dropped, as Aulén claimed, and replaced with a sociopolitical version of the evil powers. The New Testament data that support Caird's and Berkhof's theory of the Pauline doctrine of the principalities and powers become unexpected (novel) corroboration for the theory as reformulated by Yoder.[15]

1.3 Lakatosian Theology

It is important to show that existing theologies can be construed as Lakatosian research programs, because there is not much reason to propose a theological method that is entirely out of step with what theologians have been doing. But obviously there is not much reason, either, to propose a method that would make no difference at all to the practice of theology. In this section I shall indicate how I think the business of theology would go were the methodology of scientific research programs *consciously* adopted by theologians.

Up to this point I have attempted to remain theologically neutral; that is, I have attempted to discuss theological method in a way that is neutral with regard to the content of theology.[16] But there is no sharp dividing line between method and content, so my methodological recommendations will involve some theological proposals as well. The theology I sketch below may appear rather conservative in character. I do not believe that my Lakatosian method requires such an outcome. Rather, in that it is generally supposed that the more traditional theological formulations are more difficult to substantiate in light of contemporary thought, my goal here is to make my case using the more difficult material.

---

structures that perpetuate sin. It is not immediately clear, however, how to account for the further claim that Christians are freed from death, so some further elaboration is needed.

15. Unfortunately this modification would not have been acceptable to the Lundensians if it had been available to them at the time, because it conflicts with Luther's doctrine of the two kingdoms. Its acceptance would therefore violate the guidelines laid down in their positive heuristic, which I take to be the plan to treat the traditional doctrines of the church in a manner consistent with the teachings of Luther.

16. An exception is my advocacy of Edwards's theology of discernment, which I have urged on methodological grounds.

1.3.1 The Hard Core   It is conceivable that a theological research program could be built up from the bottom (i.e., by 'induction' from data). Lakatos's studies in the history of science and my own studies here in the history of theology suggest, however, that theologians need an organizing idea before they start—something like David Kelsey's "single, synoptic, imaginative judgment" regarding what Christianity is all about. The hard core, then, should contain the theologian's judgment about how to sum up the very minimum of the relevant community's faith. A Christocentric theology's core might deal only with Christ. I believe, however, that I would begin with a minimal doctrine of God, including, say, the trinitarian nature of God, God's holiness, and God's revelation in Jesus.

The function of the hard core as the source and center of a research program places constraints on its content. For instance, a core doctrine of God that emphasized God's unknowability and not much else would be impossible to relate to auxiliary theories and thus would be insupportable. Only a God who relates to and makes a difference in the world can be known through a scientific theology.

1.3.2 The Negative Heuristic   This is simply the plan or rule to protect the hard core from falsification by making additions or changes in the belt of auxiliary hypotheses. For example, with a core theory that claims God is both holy and definitively revealed in Jesus Christ, a potential falsifier appears in the conjunction of two conclusions: (1) that sexism is sinful and (2) that there is evidence in the New Testament that Jesus discriminated against women—for example, no women were included among the twelve. Now, if Jesus was sexist, then either God is not holy or Jesus is not an adequate representation of God. The negative heuristic directs the theologian not to change or abandon the hard core, but rather to seek auxiliary hypotheses that deflect the potential falsification. For example, we may add an auxiliary hypothesis regarding the character of Scripture that recognizes the effects of culture on the human authors, and adopt a reading strategy such as that of Elisabeth Schüssler Fiorenza. She posits an attempt to conceal the role of women in the Jesus movement and treats the few positive accounts as evidence for much greater involvement.[17] With this reading strategy we get a very

17. See Fiorenza, *In Memory of Her: A Feminist Theological Reconstruction of Christian Origins* (New York: Crossroad, 1983).

different picture of Jesus' relationship to women, and the hard core is saved. The addition of such a hypothesis would be a progressive move if it allows for prediction of a novel fact.

1.3.3   The Positive Heuristic   The word 'doctrine' is used in several ways in Christian thought. It may be used loosely to refer to whatever the church teaches, or more restrictively to refer to specific formulations that are given normative status relative to the various teaching functions of the church, be it the instruction of converts or the development of new 'doctrinal theology.' To avoid confusion I use 'doctrine' only in the looser sense and use 'dogma' to refer to normative statements officially adopted by various communions: *de fide* pronouncements of the popes in Catholicism, the Augsburg Confession in Lutheranism, and so on.

With this distinction at hand we can see that in many cases the dogmas of a particular communion serve the function of a positive heuristic for a doctrinal research program. That is, in such cases, the theologian's plan for development of the program is to cover all the doctrinal loci usually included in that communion's teaching, *in a way that takes account of dogmatic formulations.* Thus a contemporary theologian may elaborate a new Christological theory, not simply repeating the Chalcedonian formulas, yet governed by the heuristic rule that the limits set by Chalcedon not be violated. Likewise, the Lundensians set out to develop their entire program in ways consistent with the teachings of Luther.

Notice that this account is consistent with George Lindbeck's theory of the nature of doctrine, so long as we take him to be using the word in its more restricted sense—to apply to the normative statements of the churches rather than to their 'doctrinal' theology more generally.[18] Lindbeck describes doctrines (dogmas) as second-order discourse, as rules to guide practice and the use of first-order religious language (praise, preaching, exhortation, and the like). He points out that theology is to be consistent both with the first-order discourse of the church and with the doctrines of the church, but he does not propose a model of theology indicating how each of these boundary conditions impinges on theology. My suggestion that church dogma figures in the positive heuristic of theological research programs remedies this shortcoming.

18. In *The Nature of Doctrine* (Philadelphia: Westminster Press, 1984).

For church bodies that eschew such teaching authority, the *sola
Scriptura* principle might fulfill a similar function. The positive
heuristic would be a plan to treat all the traditional loci in a way
consistent with the teachings of Scripture. An account of *how* Scripture is to be used might be considered a part of the positive heuristic,
or it might instead be construed as a methodological auxiliary hypothesis.[19]

So we can see that the positive heuristic has an important role in
governing the character of theology. With a positive heuristic of
either sort mentioned above, a conservative or traditional theology
emerges. With a different positive heuristic (e.g., the plan to reinterpret traditional doctrines in light of existentialist philosophy) we
get a theology of a strikingly different character.

1.3.4 Auxiliary Hypotheses   As I stated above, in a systematic
theological research program there should be clusters of theories
dealing with each of the usual doctrinal topics. (Each cluster might be
a small research program in its own right.) The function of these
auxiliaries will be both to spell out the meaning of the hard core (e.g.,
to make it plain what it *means* to say that God's love is agapic) and to
provide connections between that rather abstract view of God and
appropriate kinds of data. They must be related to the core in such a
way that the vision of God summed up in the core theory helps
explain the doctrines, which in turn explain the data. In other words,
the auxiliaries must be consequences of the hard core, and the data
consequences of the auxiliaries. Explanation and confirmation being
the two sides of a single coin, the core theory is confirmed by the data
that it (indirectly) explains.

Some doctrines will be closer to the data than others. For example,
doctrines regarding the work of the Holy Spirit are very close to
accounts of the life of the church. They are little more than inductive
generalizations of observed facts. A doctrine of original sin is closely
related to (explains) the observed fact that people are affected and
infected by sin long before they reach the point of committing their
own first sin.[20]

19. See accounts of Kelsey's and McClendon's views on Scripture in Chapter 5,
sec. 7.

20. Cf. William Alston, "Christian Experience and Christian Belief," in *Faith and
Rationality: Reason and Belief in God*, ed. Alvin Plantinga and Nicholas Wolterstorff
(Notre Dame, Ind.: University of Notre Dame Press, 1983), p. 104.

On the other hand, some doctrines are far removed from Christian experience and can be confirmed only insofar as they are needed to explain other, more accessible, doctrines. For example, the trinitarian nature of God (which I have suggested as a component of the hard core, but which may instead be a separate auxiliary hypothesis) is generally thought of as a very abstract doctrine. It is easy to see, however, that it is needed to account for lower-level theories. Christians from early on have obeyed the commands of Christ and the leadings of the Spirit unconditionally—obedience owing to God alone. Accounting for this practice requires the ascription of divinity to both Christ and the Spirit—lower-level doctrinal hypotheses. But these hypotheses in turn need to be reconciled with the monotheism inherited from Israel. Christ and the Spirit must in some sense *be* God; they cannot be subsidiary deities or mere parts of God. The doctrine of the Trinity has been the traditional explanation of how Christ and the Spirit are related to Israel's JHWH. The Niceno-Constantinopolitan formulation (for some) specifies limits within which the theologian must work in attempting to solve this difficulty.[21]

Were I to produce a systematic theology, I would be sure to find a place for two important auxiliary hypotheses: first, something like Edwards's theory regarding the validity of signs of a work of the Holy Spirit. As I mentioned above, this theory legitimates the discernment practices I find so useful for producing new data for theology. It legitimates the use of judgments already reached by discerning communities, but also provides for the possibility of theological experimentation—for justifying church practices (such as obedience to Christ and the Spirit) and other phenomena as consistent with the will of God.

Second, I would include a doctrine of revelation based on Paul's account of the gifts of the Spirit and on the presumption that it was by communal discernment that the early church recognized some writings as products of the Spirit's gifts (gifts of wisdom, knowledge, prophecy)—and therefore as the word of God. Such a theory would account for the traditional claim that the Scriptures are the word of God and would thereby legitimate the use of the texts as data

---

21. See my "Theology: An Experimental Science?" in *Perspectives in Religious Studies* 15 (Fall 1988), 219–34, for a slightly amplified account.

for theology. This theory of revelation would be closely related to the Edwardsean theory used to legitimate the use of discernment.

Examination of Pannenberg's program has already made it plain that the contents of a theological research program need not be restricted to theology. I mentioned Pannenberg's inclusion of anthropology, and he has also been investigating connections between his theology, especially his doctrine of creation, and contemporary physics.[22] More on the relations between theology and science below.

1.3.5 Data   I have already made it plain that different theological programs will focus on different kinds of data. The choice will depend on the theologians' judgments about how God is made known in the world. Some will focus on revelation and will find most of their data in the Scriptures. Pannenberg's concept of God as revealed in the whole of history leads him to focus on historical facts. Others seek support for their theologies in a religious dimension of universal human experience (David Tracy and Schubert Ogden, for example). My own foci are two: Scripture and the varied results of discernment. Each of these approaches has its own problems but (not surprisingly) I see those connected with my own to be less formidable. The great task for theologians relying primarily on data from Scripture is to give an account of why the texts should be trusted to tell us about God, and not just about the Hebrews' and early Christians' *views* of God. A major problem for Pannenberg is that where he sees a historical event as an act of God, others (such as Hume) do not. He has no way to settle the issue at the level of individual perceptions, apart from invoking his entire elaborate argument for the necessity of the hypothesis of God to account for the unity of history. The religious-dimension theorists face the problem of having defined as religious, experiences quite different from those usually designated by that term. This leads one to suspect that they may have defined into existence a 'God' that has not much to do with the living God.

Answers to the question where *in general* one is to look for theolog-

22. Pannenberg, "Theological Questions to Scientists," in *The Sciences and Theology in the Twentieth Century*, ed. A. R. Peacocke (Notre Dame, Ind.: University of Notre Dame Press, 1981).

ical data do not at all rule out the possibility of making use of an assortment of other kinds of facts as well. I noted above that data for the psychologist will not be strictly psychological. Data for the astronomer will likewise be of varied sorts; for example, data from chemistry on characteristic spectral lines are important in supporting theories about the composition of the stars. In general, scientists use whatever facts come to be seen as relevant for supporting their theories, and one cannot predict what these data will be before development of the theories in question. The same will be true in theology. Facts about human behavior, ancient Near Eastern cultures, texts, languages, and many other topics may all play a role as needed in supporting a theological theory.

1.3.6 Relating Theory and Data   It is possible, now, with a survey of all the components of a theological research program before us to say more exactly how the data are to be related to the theoretical structure and how they provide confirmation for the entire program.

Carl Hempel argued that the confirmation of a theory or other hypothesis on the basis of a set of data is isomorphic with the explanation of those data by means of the hypothesis in question. That is, if a hypothesis is needed to account for a set of data, then in the absence of a better alternative, that hypothesis is thereby confirmed by those data. This is a very important insight into the nature of scientific reasoning and, in fact, into all reasoning from experience. But Hempel also claimed that hypotheses in science were to be in the form of universal generalizations, that the logical relation between the hypothesis and the data was to be deductive, and that the hypothesis should allow for the prediction of confirming observations. His work has been effectively criticized on two of these points. We often explain an observation (e.g., a crater) by means of a hypothesis concerning a singular event (e.g., a meteor collision). Furthermore, deducibility is too stringent a requirement. Even in physics, where measurements can be calculated on the basis of theories plus initial conditions, there are always missing premises (to the effect that nothing relevant has been left out of the calculation), and the calculated results turn out to be only approximations of the actual measurements. Thus we might better describe the relation between hypothesis and data as 'quasi-deductive': One must be able to produce an argument, using the hypothesis in question as a premise, which leads to the observation as its conclusion with some degree of probability.

Philosophers of science can mean either of two things when talking about *predicting* facts. Often the word has the usual, nontechnical sense—forecasting that some event will happen, that some result will be obtained by experimentation. But so long as the relation between hypotheses and data is seen as deductive, one can speak of the hypothesis as 'predicting' the fact in question in a sort of timeless sense. If today, having just observed some fact F, we show that it follows deductively from the hypothesis H, then H 'predicted' F all along, whether or not anyone made the calculations earlier.

Now, the looser the logical relation between H and F the more difficult it is to say unequivocally that F follows from H and, consequently, the more difficult it becomes to judge, when F is observed, whether H can take credit for 'predicting' it or not. Another complication is that often one can conclude from H only that a phenomenon of a certain *type* will occur. So the exact nature of the event (e.g., a particular evolutionary adaptation) cannot be predicted at all, even if it can be shown after the fact to be explainable in light of H because it is an event of a predictable type.

Thus difficulties in determining whether or not a particular fact confirms a theory are greater in some sciences than in others. They are much greater in evolutionary biology than in physics; greater in the human sciences than in biology. We may expect the difficulties in theology to be at least as great as those in the human sciences.

We have before us (in Section 1.2 of this chapter) an example of a theological research program that I claim has been confirmed by a novel fact. Let us examine the logical relations among the parts of the Lundensian program in order to judge the plausibility of that claim, using a weakened Hempelian account of confirmation combined with Lakatos's insistence that some of the confirming facts be novel. Major parts of the Lundensian program actually form a complex argument; its first premise is the hard core, and one of its conclusions is the claim that New Testament research would find an alternative account of the Pauline principalities and powers. The argument goes something like this:

(1) [God is] . . . the God of pure *agape*. (From the hard core.)
(2) The agapic nature of God's love entails that there can be no symmetrical relations between God and humankind.

(3) There are three atonement motifs in Christian history: classical, Latin, and subjective. (From historical investigation, as directed by a methodological auxiliary hypothesis.)

(4) The subjective and Latin motifs are inconsistent with (2). (This premise depends on an independent argument, whose conclusion is inserted into the main argument—in logical terminology, a lemma.)

(5) Therefore the classical atonement motif is the only acceptable one. (By elimination of all other possibilities.)

(6) The demon language of the classical motif has been found to be unacceptable. (From historical research.)

(7) Therefore the classical motif must be reformulable in a way that eliminates the demon language. (Aulén's actual prediction.)

(8) The classical motif must find appropriate evidence in Scripture. (From 5 and a methodological auxiliary regarding scriptural confirmation of Christian theology.)

(9) The standard texts for scriptural support of the classical motif are passages such as Col. 2:15, referring to the 'principalities and powers'.

(10) Therefore sound biblical scholarship must provide an alternative interpretation of the language of the principalities and powers.

We can see that this argument is by no means strictly deductive; numerous premises are missing. Nonetheless, the premises I have included provide a reasonable argument leading us to expect *something like* Caird's and Berkhof's conclusions. Notice that the actual program predicted (7)—Aulén's stated conclusion—but we can go on to 'predict' (10).

Thus the publication of Caird's and Berkhof's conclusions confirms the entire program in a weakened Hempelian sense because a statement to that effect follows 'quasi-deductively' from the hard core plus a number of additional hypotheses. It also, however, confirms the program in Lakatos's more demanding sense in that it is a novel fact: it occurred after Aulén's contribution to the program and therefore could not have been used in the formulation of that theory.

It is in treatment of data that my Lakatosian recommendations can make the greatest difference to theology. As I noted above, theologians seldom record the data upon which their theories are based. And none of them distinguish between novel facts and others. My program for the reform of theological method would lead to much greater care being taken here and to the deliberate attempt to find novel support for theological programs. In many instances research outside the library would be in order. Ideally, a theologian's own community will be a laboratory for testing the consistency of the theologian's proposals with the leading of the Spirit.[23]

2. Lakatosian Theology and Philosophy of Religion

This volume, so far, has taken up issues generally classified in three different fields: theology, philosophical theology, and philosophy of religion. One of its major theses is that the boundaries between these three areas must be transgressed. The *philosophical* problem inherited from the Enlightenment regarding evidence for theism cannot be addressed apart from the ongoing work of making sense of religious traditions—specifically in the West, from Christian *theology*—and from the task of assessing the theologians' claims to rationality (*philosophical theology*). The bulk of what has gone before being heavily theological, it is time to traverse boundaries and return to questions more philosophical.

2.1 The Evidentialist Debate in Philosophy of Religion

A new community of philosophers of religion came into being with the establishment in 1978 of the Society of Christian Philosophers—an ecumenical society both theologically and philosophically. An important challenge to the work of this volume arises from within this circle. It has been my stated assumption that in this age of agnosticism and atheism the Christian community has an obligation to provide rational support for its belief in God, if it can, in accord with the going standards of evidence.[24]

23. See my "Theology: An Experimental Science?"
24. This is a good place to make clear that I do not assume that atheists will be argued into faith. I see work such as this to be merely the removal of stumbling blocks. The cross is scandal enough without Christians adding stumbling blocks of

Alvin Plantinga argues the contrary: the mistake Christian apologists have been making since the Enlightenment has been to accept secular standards of reasoning as neutral arbiters of truth, rationality, justified belief, and to attempt to justify Christian belief on those bases.[25] The focus of Plantinga's criticism of secular theories of knowledge is foundationalism—a familiar theme from Chapter 1 above. He claims that the evidentialist challenge to theism and the conclusions of atheists such as W. K. Clifford, Antony Flew, Michael Scriven, and Bertrand Russell to the effect that the challenge has not been met are all based on "classical" foundationalist pictures of knowledge.[26] Classical foundationalism, Plantinga claims, involves resort to "basic beliefs" that are either self-evident, evident to the senses, or incorrigible. Yet this foundationalist doctrine is itself not basic in any of these senses, nor can it be made more probable on the basis of any such beliefs. Thus he concludes that it is self-referentially incoherent.

The continental Reformed tradition has regularly rejected evidentialism in the form of natural theology. This rejection, says Plantinga, can be read as an implicit rejection of classical foundationalism in favor of the view that belief in God is properly basic; the Christian has a right to believe in God without evidence or argument at all. In short, faced with a conflict between the Christian community's presumption regarding the existence and nature of God and the epistemologists' criteria for proper basicality, it is the criteria that give way in Plantinga's thought.

Now, there is something right about Plantinga's position. Christians the world over have not been waiting for the publication of books such as this one for epistemic permission to believe in God. We must take account of the strong points of Plantinga's position but, before that, let us look at its problems.

If Plantinga is correct that classical foundationalism is incoherent, why retain foundationalist vocabulary to describe the status of belief in God? Critics of foundationalism do not limit themselves to prob-

their own by failure to do the hard work of making their beliefs as clear, consistent, and rationally tenable as possible.

25. See J. Wesley Robbins, "Christian World View Philosophy and Pragmatism," *Journal of the American Academy of Religion* 51 (Fall 1988), 530.

26. See Plantinga, "Reason and Belief in God," in *Faith and Rationality*, pp. 16–93.

lems in *defining* basic beliefs; they claim there can be no such thing.[27] All beliefs are interdependent: facts in science cannot be recognized without a theoretical perspective; even simple perceptual beliefs involve a good deal of knowledge that is implicit in being able to speak a language. Plantinga comes close to recognizing this when he argues that belief in God is not *groundless*, but rather depends on fulfillment of a number of conditions.[28] But do not these conditions include acceptance of the truth of various other beliefs? Thus the successor to foundationalism is holism—the recognition that no clear distinction can be drawn between basic and nonbasic beliefs.

If the language of foundationalism is to be given up, how are we to describe the typical status of belief in God among Christians? I shall answer in two steps: first, by recalling the status of theories about God in theological research programs, and then (in the following section) by relating theological theory to the beliefs of ordinary Christians.

A persistent theme in the writings of Imre Lakatos was the denial of the adequacy of foundationalist theories of knowledge (in mathematics as well as in science).[29] His research-programs model of scientific knowledge is nonfoundationalist in its recognition that the data for scientific theories are not really data—that is, 'givens.' They are *facts* (from *facere*, to make); they are in part artifacts of the theory. Thus they undergird or support the theory only so long as certain assumptions of the research program are maintained; hence his elaborate machinery for discriminating between vicious and virtuous circularity in theory-laden support of theories.

I have suggested that in a doctrinal research program the hard core will typically contain reference to God. For example: Pannenberg's God is the all-determining reality revealed in Jesus Christ; the Lundensians' God is the God of pure *agape*. For the theologian the existence of God is presupposed (as is the existence of matter for the physicist, or of persons for the psychologist). What is at issue is a theoretical account of the nature of God (of matter, of persons). But

27. See Willard Quine's "Two Dogmas of Empiricism," discussed in Chapter 1 above; also, Richard Rorty's *Philosophy and the Mirror of Nature* (Princeton: Princeton University Press, 1980); and David Bloor's *Knowledge and Social Imagery* (London: Routledge and Kegan Paul, 1976).

28. "Reason and Belief in God," pp. 78–82.

29. See John Worrall, "Imre Lakatos (1922–1974)," in *Essays in Memory of Imre Lakatos: Boston Studies in the Philosophy of Science* 39 (1976), ed. R. S. Cohen et al., p. 3.

insofar as one's account of the nature of x is substantiated by means of the success of the entire research program, so is the assumption of the existence of x. The success of the program will be measured by its ability to predict and corroborate novel facts.

The research-programs model of theological inquiry allows us to take account of Plantinga's (correct) observation that the existence of God functions as a presupposition for Christians (theologians, at least). It also takes account of claims such as those of Barth and Thomas Torrance that the nature of the object of a science determines the science's proper methods, because the hard core influences the choice of theories of instrumentation and other theories of method.[30] But, finally, it allows one to take seriously the evidentialist challenge to theism. What may be presupposed within Christian circles is in question outside. Fortunately, the same rational structure can be both *explicatory* of Christian convictions (as seen from inside) and *justificatory* when viewed from outside. Various sorts of evidence, over the long run, do tell for or against Christian belief.

A final point at which Plantinga's approach to religious epistemology calls for criticism is his individualism. For Plantinga, knowledge (or a "noetic structure") is the property of the individual. Thus he finds it illuminating to ask whether a fourteen-year-old is obliged to have evidence for belief in God, and to answer in the negative. Yet recent epistemology recognizes that systems of belief are the property (and responsibility) of communities. Thus it may not make sense to ask each member of the Christian community to be able to give an account to the likes of a Flew or a Scriven of evidence for the existence of God, just as it makes no sense to expect the likes of me to give an account of the evidence for the existence of quarks. But that fact does not absolve the *community* from its responsibility to give an account of its beliefs to all comers—including the philosophers. For this reason I have proposed to answer questions posed to the individual theist by resorting to theological research programs—the property first of all of a community of theologians, but ultimately of the entire Christian church. Some further justification for this move is in order, however.

30. See for example Barth's *Church Dogmatics*, I/1 (Edinburgh: T. and T. Clark, 1936), sec. 6; and Torrance's *Transformation and Convergence in the Frame of Knowledge* (Grand Rapids, Mich.: Eerdmans, 1984), p. 249.

2.2 Theistic Belief and Theology

What is the relation between theologians' theories and the beliefs of ordinary Christians? Is it the same as that between the physicists' theories and a layperson's knowledge?

I suggest that theology is a *rational reconstruction* of the beliefs of a Christian community. Its job is to examine the community's belief system in order to display the relations among its parts and its justification relative to whatever else there is.[31] Theologians may reject some of the community's beliefs and modify others but must also be faithful to much of the system or else what they are doing will not be a reconstruction of that belief system.

There are important differences when we compare this case with that of the professional physicist and the amateur. In physics, nearly all knowledge comes from the professional to the amateur. In the case of theology, as here envisioned, knowledge of God begins with the amateurs (no denigration intended here—amateurs are those who participate out of love), and the professional theologian is dependent upon the findings of this community. The dependence is especially apparent when one thinks of the results of communal discernment as a primary source of data for theology. It is possible to see the proliferation of denominations and sects within Christianity as a vast array of *experiments* in Christian life and belief—in logicians' terms, attempts to produce new and more consistent models of the Christian theory.

It is tempting to identify the divergent traditions within Christianity as competing research programs; thus we might speak of the Catholic program, or the Anabaptist program. This would be a mismove, however, because we have seen that the details required for a research program involve quite specific answers to questions—such as how Scripture is to be used—about which there may be considerable disagreement among members of a single communion. Furthermore, this move would make the mistake of confusing the ends of Christianity with those of theology and science. The point of being a Christian is not first of all to acquire knowledge, but rather to be obedient to Christ. But the purpose of theology, like that of

31. For a similar account see James Wm. McClendon, Jr., and James M. Smith, *Understanding Religious Convictions* (Notre Dame, Ind.: University of Notre Dame Press, 1975).

science, broadly speaking, is knowledge. Thus it would be a mistake to think that Lakatos's methodology allows for direct competition and comparison between different Christian traditions. One can, however, compare assorted theological *reconstructions* of the beliefs of the various communions; for example, Catholic modernism versus liberal Protestantism. It is also conceivable that such comparisons be made between Christian research programs and research programs related to other religions as well. Pannenberg's plan for comparing the Christian tradition to other worldviews seems premature, however, before Christian theologians get their own house in order. At any rate, discussion of problems of religious pluralism and inter-religious dialogue would be material for another book.

2.3 Science and Lakatosian Theology

The suggestion of possible comparisons between Christian research programs and other worldviews raises the question of the relations between theology and science, because today's secular worldview is so closely allied with science. Adequate treatment of this issue would also require another book, but a few preliminary remarks may be in order.

At present there are three dominant schools of thought on the relations between theology and science.[32] One is the liberal Protestant view that religion and science are entirely different in nature; thus no interaction or conflict between theology and science is possible. Another is the conflict model, popularized by Andrew D. White's *History of the Warfare of Science with Theology*,[33] and still alive today in the creationist-evolutionist controversy. More recent than either is the attempt to show, *contra* the "two-worlds" theorists, that theology and science are similar enterprises and capable of interaction, and that, *contra* the creationists, the interaction is positive rather than negative. The most common philosophical move of those arguing for the similarity between science and theology is to propose a "critical realist" interpretation of both. That is, both science and

32. See my "Theology, the Transformer of Science: A Niebuhrian Typology for the Relation of Theology to Science," *Pacific Theological Review* 18 (Spring 1985), 16–23, reprinted in *The Center for Theology and the Natural Sciences Bulletin* 5 (Fall 1985).

33. New York, 1896.

theology are said to give us tentative, approximative pictures of reality.[34] There are two difficulties with this approach: First, critical realism is a problematic position philosophically—it is difficult to get clear about what it could mean unless one interprets it either as a truism or as an outrageous claim to have some knowledge of reality apart from our ordinary human ways of knowing.[35] Second, even if critical realism stands up to philosophical scrutiny, it is not clear how it solves the problem for which it is invoked, namely to give an account of how theology and science interact. One is left with two equally valid but complementary pictures of reality—a situation that comes close, after all the struggle, to falling back into a version of the two-worlds approach.

The present work contributes to the third, "interactionist," position in that it sets out to show plainly that (potentially at least) theology is methodologically indistinguishable from the sciences. Furthermore, it opens the door to a very straightforward account of interaction. We can envision at least two sorts:

(1) Theology-and-science hybrids: Research programs are networks of theories with assorted kinds of data. Incorporation of auxiliary hypotheses respects no conventional boundaries between different sciences. For example, theories from the science of optics play a role as theories of instrumentation in both astronomy and microbiology. Thus there is no reason in principle why theories originating in theology cannot be included as auxiliary hypotheses in scientific research programs, and vice versa.[36] As a matter of fact, scientific hypotheses occur regularly in theology—for example, as aids in the interpretation of biblical texts.[37]

---

34. See, for example, Ian Barbour, *Myths, Models, and Paradigms* (New York: Harper and Row, 1974); Arthur Peacocke, *Intimations of Reality: Critical Realism in Science and Religion* (Notre Dame, Ind.: University of Notre Dame Press, 1985).

35. See my "Relating Theology and Science in a Postmodern Age," *The Center for Theology and the Natural Sciences Bulletin* 7 (Fall 1987), 1–10; "Scientific Realism and Postmodern Philosophy," *British Journal for the Philosophy of Science* 41 (June 1990); and "From Critical Realism to a Methodological Approach: Response to Robbins, van Huyssteen, and Hefner," *Zygon* 23 (September 1988), 287–90; and J. Wesley Robbins, "Seriously, but Not Literally," in that same issue, pp. 229–46.

36. See my "Acceptability Criteria for Work in Theology and Science," *Zygon* 22 (September 1987), 279–98.

37. For example, there are the scientific theories involved in dating ancient texts and sociological theories used in interpretation.

Lakatos recognized that metaphysical views of reality often provide the hard core of scientific research programs. Metaphysics and theology are old bedfellows so, for instance, some historians have pointed out that the concept of inertia included in Newton's hard core was motivated by his Calvinist theology.[38] Pannenberg's program, as I would reconstruct it, has a metaphysical-theological hard core and incorporates theories from physics and anthropology (and, potentially, other sciences) as auxiliary hypotheses. So the notion of a hybrid theological-scientific program is not at all far-fetched.

The proscription of the mixing of science and theology began in the modern era as a way to preserve theology from corrupting influences of "natural philosophy." Later—for example, among the logical positivists—the motive was to preserve science from the corrupting influences of theology and metaphysics. With Lakatos's work the proscription of metaphysics is seen to be incompatible with the typical development of scientific programs. If my arguments are sound, Lakatos's proposal for distinguishing science from theology fails, and the burden of providing a new principle of demarcation shifts to those who would exclude theology from interaction with the other sciences.

(2) A second possibility for interaction is competition between theological and scientific research programs. An extremely important area for theologians to attend to is the competition between theological accounts of religious experience and those of the secular sociologists and anthropologists. If the human scientists can provide more progressive accounts of religious behavior and experience than theologians can on their home ground, that is bad news indeed for theists!

3. Postmodern Philosophical Theology

Recently it has become fashionable for philosophers, literary critics, architects, and even theologians to call their work postmodern. Some of these uses are incompatible. In this section I shall propose a convention for use of the word in philosophy and theology. Having done so, I shall be able to place several of the claimants for the title

38. See Eugene Klaaren, *Religious Origins of Modern Science* (Grand Rapids, Mich.: Eerdmans, 1977).

'postmodern' and will also be in a better position to reconcile my
work with that of philosophical theologians whom I judge to merit
the title. It will turn out that the works of George Lindbeck and
Ronald Thiemann are indeed postmodern. Thiemann and Lindbeck
see the end of foundationalism as a significant fact for religious
epistemology. Some take them to have concluded from this fact that
the enterprise of Christian theology (as a whole) needs no justification. If such a reading is correct, then we meet a serious challenge to
the present book: is all this really necessary?

3.1 Modern Thought

The beginning of the modern period has been characterized as the
"turn to the subject." It was the era in which epistemology became
the center of philosophy, replacing cosmology and metaphysics—at
least until philosophy took the (still modern) "linguistic turn." It was
the age of skepticism, reductionism, individualism, and the "flight
from [traditional] authority."

These near-truisms point to three central philosophical theses that
dominated modern thought. The first was epistemological foundationalism. The second was the representational or referential theory
of language—the view that language must gain its primary meaning
by representing the objects or facts to which it refers.[39] The third
pillar of modern thought was atomism or reductionism, exemplified, for instance, by modern individualism—an approach to ethics
and political philosophy that saw the individual as prior to the community.

Just as atheists tend to differ depending on the kind of theism they
reject, the typical minority positions of the modern period were
shaped by these dominant themes above. Hence modern *skepticism*
appeared in the form of doubt about the foundations and about
whether ideas or language represent and how accurately they do so.
Expressivist theories of language describe religious or ethical discourse,
in the absence of reference, as mere expression. Modern *collectivism*
presupposes individualism by viewing societies as mere aggregates
of like individuals.

39. Perhaps the classic exposition of this position is in Ludwig Wittgenstein's
Tractatus Logico-philosophicus (London, 1922).

Dominant position versus typical counterposition suggests that modern thinkers might be represented as falling along axes—for example, the logical positivists with high hopes for reconstruction of science would be placed at the extreme foundationalist end of an epistemological axis, with the somewhat chastened neopositivists a small distance toward skepticism.

There were close interrelations among modern positions, seen most explicitly, perhaps, in logical atomism, where *atomic* sentences were linguistic *representations* of *foundational* facts. Thus, extending the metaphor, we may imagine a three-dimensional conceptual 'space,' in which to locate modern thinkers by means of 'Cartesian coordinates.'

3.2 Postmodern Theology

Once modern philosophy is thus characterized, it becomes clear that recent philosophy (during the past thirty to forty years) has made dramatic shifts in emphasis—drastic enough to warrant the claim that modern philosophy has come to an end. James McClendon and I have argued that the term 'postmodern' should be reserved for philosophers who have removed themselves decisively from modern conceptual 'space' without reverting to premodern modes of thought.[40] Postmodern theologians, then, will be those whose philosophical presuppositions are postmodern rather than modern. We see revisionary philosophical moves in three areas. The first is the change from foundationalism to holism in epistemology. The second is the change from the modern emphasis on reference and representation in philosophy of language to J. L. Austin's and Ludwig Wittgenstein's emphases on language as action, and meaning as use.[41] Third, we see a major postmodern shift in Alasdair MacIntyre's, Robert Bellah's, and others' renewed sense of the importance and irreducibility of community.[42]

40. See Murphy and James Wm. McClendon, Jr., "Distinguishing Modern and Postmodern Theologies," *Modern Theology* 5 (April 1989), 145–68.

41. See Wittgenstein's *Philosophical Investigations* (New York: Macmillan, 1953); and Austin's *Philosophical Papers* (Oxford: Clarendon Press, 1961).

42. See MacIntyre's *After Virtue*, 2d ed. (Notre Dame, Ind.: University of Notre Dame Press, 1984); and Bellah et al., *Habits of the Heart* (Berkeley: University of California Press, 1985). A more nuanced account of the shift from modern to

Just as there are connections among modern philosophical positions, so too are there affinities among their postmodern replace-
ments. Both holism in epistemology and the theory of meaning as
use cut against the individualism of the modern period. Moderns
assumed that any individual (given the basic sensory and intellectual
equipment) was as competent as any other to form justified beliefs
and to speak the language. Society's knowledge and language were
merely the collection of the individuals'. In postmodern thought,
however, the community plays an indispensable role. It is the community of scientists which decides when to take anomalous facts
seriously. The community must decide where to make changes in the
Quinian net of beliefs. The conventions, the language games, in
which one participates precede individual speech and determine what
can and cannot be said. In short, language and the search for knowledge are practices dependent upon tradition—they are communal
achievements. Thus escape from either of the modern epistemological or linguistic axes calls for a corresponding detachment from the
individualist axis as well.

The postmodern positions of holism in epistemology and meaning
as use are themselves closely related. In adjusting our network of
beliefs to fit the world, we may change particular beliefs about the
world, but we may also adjust meanings in order to restore consistency. Belief and meaning are inextricably related.

On this view of the matter, Lindbeck's "post-liberal" theology
with its holist theory of knowledge and its recognition of the diverse
functions of language is clearly postmodern.[43] So, too, is Thiemann's closely related proposal for a nonfoundational justification of
the doctrine of revelation.[44] But the deconstructionist-inspired theology of Mark C. Taylor might better be seen as the last gasp of
modernity, with its foundationalist-style skepticism and reliance on

---

postmodern modes of thought would require a consideration of Hegel's contributions. Because of his antifoundationalism and nonreductive sense of community, he
may be an important founder or precursor of postmodernity.

43. In *The Nature of Doctrine*.

44. See his *Revelation and Theology* (Notre Dame, Ind.: University of Notre Dame
Press, 1985). Another genuinely postmodern work is McClendon and Smith's *Understanding Religious Convictions*. For a somewhat similar point of view, see David
Griffin, *The Reenchantment of Science: Postmodern Proposals* (New York: SUNY Press,
1988).

referential theories of language to substantiate his claims of mean-
inglessness.[45] Thus it is only the works of Lindbeck and Thiemann
that will concern us here—they are close philosophical kin.

In his proposal for a nonfoundationalist approach to theology,
Thiemann makes use of Quine's model of knowledge as a net or
web, where there is no foundation or starting point, and justification
of a problematic belief involves showing its connections with beliefs
held to be unproblematic. Thiemann conceives of nonfoundationalist theology as primarily descriptive of Christian belief and practice
(in contrast to explanation or theoretical defense). It attempts to
show the "intelligibility, aptness, and warranted assertibility" of
Christian beliefs, paying close attention to the patterns inherent in
particular beliefs and practices rather than to a general theory of
rationality. That is, it seeks its criteria of judgment within the first-
order language of church practice, evaluating and criticizing beliefs
according to criteria internal to Christian faith. Thiemann holds the
following to be characteristics of a nonfoundational theology: (1)
justification of belief is specific to Christian faith, community, and
tradition; (2) the second-order language of theology is closely tied to
the first-order expressions of the community's faith; and (3) theology
employs holist justification, which seeks the relation between a dis-
puted belief and the web of interrelated beliefs within which it rests—
a process of rational persuasion. Lindbeck advocates a similar account of "intratextual" theology.

So far, justification of theological claims amounts to showing that
they are consistent with the rest of the system; no attention is given to
the question why accept the system as a whole? At this point one
might hear Jeffrey Stout's question echoing in the back of one's
mind: "Are the truth of a group's beliefs and the validity of their
reasoning simply up to them?" Thiemann offers no guidance at all
for choices between competing theological systems. In Lindbeck's
work, however, the question of truth arises in two ways. One is that
of the consistency or coherence of each part of the system with the
rest—first-order propositions, second-order theological and doctrinal statements, and practices of the community. Second, Lindbeck
raises the question about the 'truth' of the religion itself. But this is

45. In Taylor, *Erring: A Postmodern A/Theology* (Chicago: University of Chicago
Press, 1984).

better expressed, he says, as a question about the *adequacy* of the system as a whole to conform its adherents in the various dimensions of their existence to "the ultimate reality."[46] In addressing epistemological questions, Lindbeck compares religions to comprehensive scientific theories (paradigms). Within a theory (paradigm, religion) one evaluates truth claims on the basis of consistency among beliefs and agreement with experience. In evaluating a theory (paradigm, religion) as a whole, one uses a variety of criteria, including its ability to assimilate new data and provide an intelligible interpretation of varied situations.[47]

Lindbeck has been accused of offering a cultural-relativist view of religious belief, but I think we can see that this charge is unfair. We might justly conclude, however, that his account of how one justifies claims regarding the truth of a religion is inadequate. The benefits that my approach to theological methodology offers over Lindbeck's are exactly the benefits that Lakatos's methodology offers scientists over Kuhn's: the criterion of empirical progress offers a way to arbitrate between two or more equally consistent, comprehensive views of reality.

A second worry arises from Thiemann's work. He has criticized modern approaches that conceive of theology as a theoretical activity like any other "seeking the goods associated with theoretical activities."

> Theology takes its place beside other theoretical inquiries ruled by general principles of rationality. Whenever a theoretical inquiry diverges from those general principles, it must justify that divergence by a special apologetic argument. Since Christian claims often appear exceptional when judged by general principles of rationality, theology must regularly undertake to defend its divergent tendencies. If a foundationalist theologian is concerned at all to guard the distinctiveness of the theological subject matter, apologetics inevitably emerges as the *primary* theological task. The two most important theological activities—the development of a universal justificatory argument and the defense of Christian claims before the bar of rationality—are carried on independent of the internal logic of Christian belief and practice

---

46. *The Nature of Doctrine*, pp. 64–66.

47. I believe it is a category mistake to compare a religion to a theory. See above, sec. 2.2.

and with little reference to criteria of judgment internal to the Christian tradition.[48]

In this way Thiemann implicitly raises the question whether rational adjudication between differing theological systems using a Lakatosian methodology is not a case of invoking "general principles of rationality" that do violence to the internal logic of those systems.

It may be helpful to recall at this point that Lakatos developed his theory of scientific method to account for choices between competing approaches to scientific research. He saw in Kuhn's account of science a threat to cherished views of scientific rationality: How could the choice of one program over another be based on rational considerations if the data, theories of instrumentation, methods of research, and judgments about the significance of solutions to problems were all "paradigm dependent"? Lakatos's solution was to take a set of desiderata for scientific theories—empirical fit, heuristic power, avoidance of 'ad hocery'—and derive from their joint demands a single criterion for assessing the "progress" of a paradigm or research program. Furthermore, he and his followers have shown that this criterion—the demand for novel corroboration—is in no way alien to the practices of science. In fact it accounts for scientists' judgments in choosing one program over another *in a variety of fields*. And it interferes not at all with the standards of research (laboratory procedures, standards for problem solutions, and so on) inherent in the various paradigms.

In Chapter 4 and earlier in this chapter, I presented examples of theological programs, showing that Lakatos's requirements for a scientific research program are not only not alien to theology, but do in fact serve as a useful device for promoting understanding of the theoretical development of theological programs. In Chapter 5 I developed a proposal to use the results of discernment as a source of data for theology. This proposal is exactly in line with Thiemann's concern to employ "criteria of judgment internal to the Christian tradition." Lakatos's methodology interferes no more with these internal standards of judgment than it does with, say, the chemist's laboratory procedures.

Thus we can see that the Lakatosian methodology I propose for

48. *Revelation and Theology*, p. 74.

theology reconciles the desire for standards to assess the relative merits of competing belief systems with Thiemann's warning against the imposition of general principles of rationality that ignore or distort the patterns of justification and evaluation inherent in the first-order language of Christian belief and practice. Each theological research program can display its own internal logic, based on the convictions and justifications offered by the relevant community. Given such display of the internal rationality of a number of Christian conviction sets, the Lakatosian distinction between progressive and degenerating programs becomes available as a means of rational adjudication between systems—adjudication by means of standards recognizable by all.

It must be noted that one could expect the methodology advocated here to apply only to western theologies constructed since the rise of empiricism. The methodology formulates intuitions about knowledge and rationality which may be widely shared only in our culture: that knowledge must relate to experience, that knowledge grows and develops, and that the discovery of new information is at least as important as maintaining consistency within the belief system. Insofar as these basic attitudes regarding knowledge are not shared, for example, by Tyrrell's scholastic opponents, the use of this methodology would indeed constitute the imposition of an alien standard of rationality.

On the basis of the foregoing considerations I conclude that a nonfoundationalist approach to theology guided by current philosophy of science is indeed possible. Its subject matter will be God and God's relations to the worlds of science and history and human interiority. But because theology seeks to assess the "aptness and warranted assertability" of Christian claims (Thiemann), responding in many ways to many sorts of empirical data, it finds itself to be one science among many. Therefore it will employ the best theory of scientific method available, along with the particular concrete methods suited to its own subject matter—just as all sciences have their own concrete experimental or observational methods but share forms of reasoning. Lakatos's theory of scientific reasoning has been shown to apply to sciences as diverse as economics, high-energy physics, and evolutionary biology; now we may add theology to the list. Use of forms of reasoning common to all kinds of science provides the basis for persuading those outside of the Christian community of the rationality and intelligibility of its claims.

It must be noted, however, that the theological methodology proposed here, while providing a counter to total relativism, offers no absolutist view of rationality, even within our (loosely) empiricist worldview. There is no instant rationality in science or theology. Programs take time to develop. There are sometimes no noticeable differences in the degree of progress exhibited by competitors. Judgments to reject one program in favor of another may be made too hastily, because programs sometimes degenerate for a time but later turn around and become progressive. Thus we have no guarantee that the One True Theology will be found amid today's pluralism. Still, I believe I have offered a better starting point for discussion of the merits of competing theologies than anyone else to date.

When we consider the application of Lakatos's methodology to theology rather than to science, some of its weaknesses become less problematic. Recall Feyerabend's objection (in Chapter 3 above) that Lakatos's methodology is empty without a time limit on degenerating programs, yet with such a limit it is likely to force the premature abandonment of important scientific programs. Theology and science differ, though, in that theological programs tend to proliferate in ways that scientific programs cannot. Theologians often commit themselves to their theologies not only as scientists to pet theories, but also as believers to their faith. Theology does not require the same financial resources as does much scientific research, so competition between theological positions does not become a life and death struggle for funding. In a setting such as this, Lakatos's methodology is unlikely to result in the premature abandonment of programs (as Feyerabend fears it will in science), but will instead result in more serious attempts to exhibit the coherence and support of positions to which theologians are antecedently committed, and will therefore contribute to the clarity of theology generally and to the discovery of new theological facts. In short, application of Lakatos's criteria might provide much-needed pressure to make theological programs perform.

A second major difference between science and theology is that few today question the rationality of science itself (Feyerabend being one notable exception),[49] whereas many not only question the rationality of theology but are positively convinced that it is irrational

49. See Feyerabend, *Science in a Free Society* (London: New Left Books, 1978), chap. 1.

in all its forms. Thus in philosophy of science it is important to haggle over fine points (such as when it became irrational to pursue the phlogiston program). In philosophy of religion the important point of contention is still whether it is *possible* to be a rational theologian. Here the game is won by anyone who can show that theology is in the same ball park with science, and no points should be taken off if one cannot give sharp answers about when to give up on Pannenberg's or the modernists' programs. As Alan Musgrave points out (speaking of science), those involved in the program have a pretty good sense of whether there is still productive work to be done there or not.[50] At this point I would say that the field of theology is wide open. No program has been actively pursued with Lakatos's guidelines in mind. And there's no telling what might happen were this to be done.

4. Retrospect: Hume and Stout Revisited

In Chapter 1 I set the challenge for the present work by reviewing David Hume's skeptical writings on religion. In Hume's day a sharp distinction was drawn between natural theology, newly reconceived by means of probable reasoning on the basis of empirical evidence, and revelation theology, based on a foundationalist doctrine of revelation, buttressed by appeal to miracles and fulfilled prophecies. Jeffrey Stout speaks for many philosophers today who believe that Hume has permanently demolished the rational credibility of theology. But the empiricist tradition has gone its own way since then, providing important new insights for the understanding of empirical knowledge. So the challenge was to find a theology commensurable with *current* canons of probable (scientific) reasoning.

In Chapter 2 I turned to the theology of Wolfhart Pannenberg because his work displays a combination of three important ingredients for a response to the challenge. First, he repudiated the distinction between natural and revelation theology. His particular strategy was to assimilate revelation to history, making the scriptural texts data of the same sort as any other text and demanding confirmation

50. See "Method or Madness?" in *Essays in Memory of Imre Lakatos*, ed. Cohen et al., pp. 457–91.

of the religious tradition's account of the "all-determining reality" on the basis of an assortment of facts and observations. His second major contribution was a call for the adoption of scientific method for theology. His particular approach was to devise a theory of scientific method that assimilates scientific reasoning to hermeneutics. His third contribution was to insist that Christian theology be the science of God (not the science of the Christian religion), while theological knowledge of God was to be attained *by means of* critical evaluation of the Christian tradition.

I have proposed a view of theology that parallels Pannenberg's in intent in all three of these respects, but differs in its execution. Earlier in this chapter I advocated a view of theology as the science of God. In Chapter 5 I gave special attention to Christian discernment because it is an important epistemic practice that yields knowledge of God's involvement in the life of the church.

In Chapter 3 I recommended the scientific methodology of Imre Lakatos as a more sophisticated account of the practice of science than that used by Pannenberg. In Chapters 4 and 6 I showed that this methodology serves to describe theology, and I have suggested that as a norm for evaluation of theological programs it *works* better than Pannenberg's. The latter calls for evaluation of theological programs on the basis of how well they manage to account for the totality of history. Lakatos's methodology calls for evaluation on the basis of comparative progress of the programs. We know from the history of science that it is often possible to assess the relative progress of competing programs.

The distinction between natural and revelation theology disappears in my proposal as well. In Chapter 5 I suggested that in a research-programs theology, doctrines of revelation no longer serve as the foundation upon which systematic theology rests or as part of theological prolegomena, but are instead included in the theoretical structure to explain communities' assorted ways of "taking Scripture as authority." Thus Scripture is likely to figure prominently in any Christian research program for several reasons, but the texts themselves are not required to provide a category of privileged or indubitable data. The justification for a community's use of Scripture (its doctrine of revelation or some other doctrine) will arise from the internal form of the community rather than from special apologetic moves by philosophical theologians and will function as an auxiliary hypothesis in its doctrinal research programs.

I suggested in Chapter 2 that Hume's agnostic arguments were not motivated simply by antireligious sentiment but were part of a comprehensive program to counter "false metaphysics"—the alienation of reason from the language and customs of the common life that give it meaning. Hume saw the prophetic-providential view of history as the cause of the alienation of judgments regarding politics and morality from their basis in custom, because it led to the view that political history was driven by the plans of a God who stood outside it, and the view that morality was enforced by means of expectations of eternal reward or punishment. He attacked this view of history by showing that it could not be supported by the new probable or experimental reasoning.

I have now proposed a view of theology compatible with recent developments in the tradition of probable reasoning. It is not accidental, I believe, that the recent turn in theology from foundationalist to nonfoundationalist theorizing accords exactly with Hume's positive views on the proper relation of reason to custom—"philosophical decisions are nothing but the reflections of common life, methodized and corrected."[51] Nonfoundationalist theology is not alienated from the common life of the church by the dictates of an autonomous reason; rather, reason is assigned the more modest task of reforming the language and practices that have been built up over the years, bringing to explicit awareness the standards implicit in its conventions and, perhaps, arbitrating the claims of competing forms of life. I therefore conclude that the theology proposed in these pages would in principle be satisfactory to Hume were he alive today,[52] because the nonfoundationalist philosophy of science of Lakatos et al. bears a close relation to the ongoing common life of scientific practice—bringing to explicit awareness the standards implicit in its judgments on theories.[53]

So Stout's prognosis for theology after Hume was overly pessimistic. He claimed that the end of epistemological foundationalism

51. *An Enquiry concerning Human Understanding*, ed. L. A. Selby-Bigge (Oxford: Clarendon Press, 1902), p. 162.

52. At least the theology would be acceptable to Hume if the science were; his arguments cut just as sharply against science as against theology.

53. Not all philosophers would agree, however; Paul Feyerabend believes that all philosophy of science, even that of his friend Lakatos, serves a rationalist tradition alienated from its roots in the common life. See *Science in a Free Society*.

made no difference to the effectiveness of Hume's arguments; in fact it has made a significant difference. In short, Stout failed to take account of the capacity of theology to adapt to the new nonfoundationalist epistemology best represented in Lakatos's historicist theory of scientific method.

In sum, two interrelated goals have shaped the present volume. One was to provide an answer to modern philosophical skepticism regarding claims to knowledge of God. The other was to provide theologians with a clear criterion for evaluating theological proposals—Lakatos's "progress" via "novel facts." Insofar as that criterion is fulfilled by theological theories, those theories are shown to be scientific and therefore acceptable in the age when science lays down "where and in what sense we may speak of knowledge."

Index

Library of Congress Cataloging-in-Publication Data

Murphy, Nancey C.
    Theology in the age of scientific reasoning / Nancey Murphy.
        p.   cm. — (Cornell studies in the philosophy of religion)
    ISBN 0-8014-2400-3 (alk. paper)
        1. Theology—Methodology,   2. Religion and science—1946–
I. Title.   II. Series.
BR118.M88   1990
230'.01—dc20   89-39375

Cornell Studies in the Philosophy of Religion

EDITED BY WILLIAM P. ALSTON